A BASIC GUIDE TO POWER ELECTRONICS

A BASIC GUIDE TO POWER ELECTRONICS

Albert Kloss
BBC, Brown, Boveri & Company Ltd.,
Baden, Switzerland

Translated by
K. G. King

A Wiley–Interscience Publication

JOHN WILEY & SONS
Chichester · New York · Toronto · Brisbane · Singapore

First published in German as
Leistungselektronik ohne Ballast
by Franzis-Verlag GmbH, München, 1980

Library of Congress Cataloging in Publication Data:
Kloss, Albert.
A basic guide to power electronics.

Translation of: Leistungselektronik ohne Ballast.
'A Wiley–Interscience publication.'
Includes index.
1. Power electronics. I. Title.
TK7881.15.K55 1984 621.3815 84–7398

ISBN 0 471 90432 5

British Library Cataloguing in Publication Data:
Kloss, Albert
A basic guide to power electronics.
1. Electric current converters 2. Power electronics
I. Title II. Leistungselektronik ohne Ballast. *English*
621.3815′322 TK7872.C8

ISBN 0 471 90432 5

Phototypeset by Mathematical Composition Setters Ltd, 7 Ivy St, Salisbury, Wilts.
Printed by St. Edmundsbury Press Ltd., Suffolk.

Contents

List of Symbols

CURRENTS

Instantaneous values

i	Alternating current, transformer current, valve current
$i_{\sim}$	Alternating current, supply (source) current
$i_{R,S,T}$	Phase currents
i_k	Short-circuit current, commutating current
i_{Th}	Thyristor current
$i_{1,2,\ldots,}$	Valve currents
i_1	Fundamental current
i_g	Triggering (gate) pulse
i_E	D.C. voltage-source current
i_d	Direct current
i_R	Current in resistor
i_C	Current in capacitor
i_m	Maximum value
I_m	Amplitude
I_F	Foward current, maximum value
i_{rr}	Reverse current peak

Relative values

i_d	Direct current

Mean values

I_d	Direct current
I_{do}	Direct current with $\alpha = 0^{\circ}$
I_{dN}	Rated (nominal) direct current
I_{dk}	Short-circuit direct current
I_{VAV}	Valve current
I_{ThAV}	Thyristor current

R.M.S. values

I	Alternating current, transformer current
I_N	Rated (nominal) transformer current
I_k	Short-circuit current

I_{Vrms}	Valve current
I_{Thrms}	Thyristor current
I_1	Fundamental component
I_n	Harmonic current
I_s	Supply current

VOLTAGES

Instantaneous values

v	Alternating voltage, line voltage, valve voltage
$v_{R,S,T}$	Phase voltages
v_{ph}	Phase voltage
v_n	Harmonic voltage
v_s	Supply voltage
v_{sec}	Secondary voltage
v_V	Valve voltage
$v_{1,2,\ldots}$	Valve voltage
v_{Th}	Thyristor voltage
$v_{L,Lk}$	Voltage across inductance
v_C	Capacitor voltage
v_d	Direct voltage
v_D	Diode voltage
v_{rr}	Reverse recovery voltage peak
V_{Vm}	Valve voltage, maximum value

Relative values

v_d	Direct voltage
dx	Inductive d.c. voltage drop
dr	Ohmic d.c. voltage drop
e_k	Total short-circuit voltage
e_{kt}	Transformer short-circuit voltage

Mean values

V_d	Direct voltage
V_{dio}	Ideal no-load direct voltage, $\alpha = 0°$
V_{do}	Direct voltage with $\alpha = 0°$
E	D.C. voltage source
A	Voltage–time integral (area)

R.M.S. values

V	Line voltage
V_1	Fundamental component
V_{Vrms}	Valve voltage
V_{ph}	Phase voltage

V_{sec}	Secondary voltage
V_p	Primary voltage
V_0	Source voltage, no-load voltage

POWERS

P	Active power (mean)
S	Total apparent power
S_1	Fundamental frequency apparent power
S_t	Transformer rated power (VA)
Q	Total non-active power (reactive power)
Q_1	Fundamental frequency reactive power
Q_H	Harmonic non-active power, distortion power
Q_T	Transformer short-circuit volt-amperes
Q_s	Supply short-circuit capacity
p	Relative active power, instantaneous power
q_1	Relative fundamental frequency reactive power
q_C	Relative capacitive reactive power
s_1	Relative fundamental frequency apparent power
W	Energy

PARAMETERS, ABBREVIATIONS

L	Inductance in d.c. circuit
L_k	Commutating inductance
C	Capacitance
R	Resistance
R_i	Equivalent resistance
x_k	Commutating reactance
V	Valve
D	Diode
L_F	Filter inductance
L_μ	Transformer no-load (magnetising) inductance
K	Cathode
A	Anode
p,n	Semiconductor conductivity types
REC	Rectifier
INV	Inverter
G	Generator
M	Motor
ASM	Asynchronous (induction) motor
SM	Synchronous machine
HVDC	High-voltage d.c. (transmission)
Hg	Mercury
Si	Silicon
Se	Selenium
n	Harmonic order, rotational speed

ANGLES, ANGULAR INTERVALS, FACTORS

α	Delay angle
α'	Natural delay angle
ϕ	Phase angle
ϕ_1	Phase angle of fundamental component
ωt	Angular interval at supply frequency
ω	Angular frequency of alternating (supply) voltage
ω_0	Natural angular frequency of oscillatory circuit
λ	Conduction angle, power factor
μ	Overlap (commutation) angle
f	Frequency
T	Period
τ	Time constant

HEAT

R_{th}	Thermal resistance (steady state)
r_{th}	Transient thermal resistance
P	Loss in valve
v_0	Threshold voltage
r	Equivalent (incremental) resistance
θ	Temperature
k_f	Current form factor

Preface

This book is directed both to students and to practising engineers and technicians.

The pillars of the book are the diagrams. They carry the most important information and it is possible as a first step to use the work as a picture book.

The main emphasis is placed upon a physical understanding of the problems considered. Practically oriented mathematical relationships are associated with, but separated from, individual sections of text. The reader can study the book in its essentials without having to become involved with mathematics.

The book is intended to serve as a textbook, with systematically structured material, and also as a practical work of reference.

The discussion concerns the most important aspect of power electronics—converter circuits for three-phase systems—and is centred on the three-phase bridge circuits with natural commutation which are the most widely used. Systematic treatment of the fundamentals is supported by the latest findings from industry and research. Many of the diagrams are based upon computer calculations.

The author has drawn upon long experience as an engineer with Brown Boveri and Co. at Baden, Switzerland, since 1969, and as delegate laecture at the Winterthur Technical and Engineering College (1971–1983).

The author would like to thank Mrs. M. Lochner for the painstakingly drawn diagrams and Mr. G. Lochner (Dipl. Ing.) for his critical checking of the manuscript.

IMPORTANT NOTICE

The circuits and methods reproduced in this book are given without any consideration to the patent situation. They are chosen purely for the purposes of interest and instruction, and may not be used commercially.

All the circuits and technical data in this book have been collected and assembled by the author with the utmost care and reproduced under the closest control. Nevertheless, errors may not have been entirely avoided. The publishers feel obliged, therefore, to make it clear that they can offer no guarantee, nor accept any legal responsibility or liability, in regard to any consequences attributable to erroneous statements. The author and the publishers will be grateful if any errors are brought to their notice.

1

The Structure of Power Electronics

FUNDAMENTALS

The principles of power electronics cover electrical circuits and apparatus which achieve the control or conversion of electrical energy by means of electronic valves.*

Power electronic equipments are generally referred to as 'converters', and the valves are accordingly termed 'converter valves'.

Valve action takes place in solid-state semiconductors. Because power electronic apparatus contains no moving parts, it is often referred to as 'static' equipment. The term 'power electronics' stems from the development of high-power semiconductor valves during the nineteen-sixties. The expressions converter or rectifier technology were previously used.

CONVERTERS

Converters constitute the hardware of power electronics. According to their basic functions, converters can be divided into the following groups:

Rectifiers (a.c.–d.c. converters)
Inverters (d.c.–a.c. converters)
A.C.–a.c. converters
D.C.–d.c. converters

Rectifiers convert alternating current (a.c.) into direct current (d.c.). Conversely, inverters convert direct to alternating current.

With converters that convert from a.c. to a.c. or from d.c. to d.c. it is possible to change voltage, frequency, phase number, or polarity.

Circuits that can function as rectifiers or inverters, according to the way in which they are controlled, are often referred to simply as 'converters'.

With the exception of rectifiers, which can operate in an uncontrolled mode, all converters embody controllable valves.

A.C.–a.c. and D.C.–d.c. CONVERTERS

This group includes the following subgroups:

Direct a.c. converters (direct frequency changers, cycloconverters)

* As often in continental usage, and in accordance with the *IEC International Electrotechnical Vocabulary*, Chapter 551, the term 'valve' is applied generally in this book to the class of rectifying and switching devices which includes solid-state diodes and thyristors as well as vacuum, gas, or vapour-filled tubes.

Indirect a.c. converters (d.c.–link frequency changers)
D.C. regulators (d.c. choppers)
A.C. regulators

The direct frequency changer converts alternating current of one voltage, frequency, and phase number into a different voltage, frequency, and phase number directly, without a means of energy storage. The output frequency is usually lower than the input frequency.

The d.c.–link frequency changer serves basically the same purpose as the direct frequency changer, but includes in addition a means of energy storage in a direct current or direct voltage intermediate circuit. The output frequency is consequently independent of the input frequency.

In both d.c.–link and direct frequency changers energy can be translated in either direction. The d.c.–link frequency changer feeding a synchronous machine constitutes the so-called converter motor; connected in the rotor circuit of an asynchronous machine it forms a subsynchronous converter

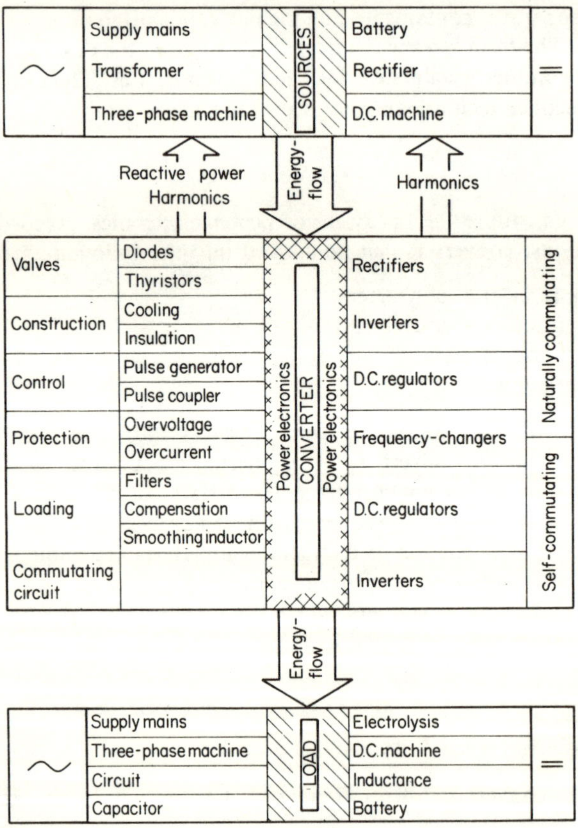

Figure 1. Diagrammatic presentation of the role of converters in energy conversion and control. The converters lie in the path of energy flowing between the source and the load. On the left are listed the components and the problem areas, on the right the classification of converter systems

cascade (the static Kraemer system), and in electricity supply systems it is used for high-voltage d.c. transmission.

The d.c. regulator facilitates the variation of d.c. voltage, without any intermediate circuit, but normally by dint of a forced commutation circuit (or valves with a turn-off facility).

The a.c. regulator affords a variation of a.c. voltage, at an unchanged frequency, again without an intermediate circuit.

The two regulators are not always counted as 'converters' but may be regarded as a distinct group. Figures 1 and 2 illustrate the classification of converters in terms of power flow, components, and grouping.

INTERNAL OPERATION OF CONVERTERS

Converter circuits are assemblies of converter valves, which function as controlled switching devices. The required conversion and control of voltage

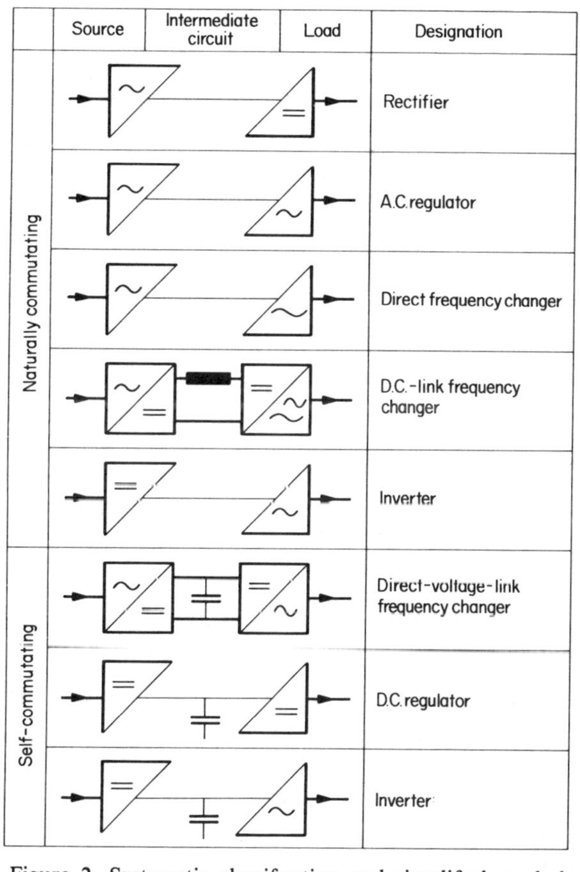

Figure 2. Systematic classification and simplified symbols of converter types. Converter circuits fall into two main groups: converters with natural (external) commutation and converters with forced commutation (self-commutating)

and current are achieved by the appropriately directed turning on and off of the valves. Only low-power control signals are needed to initiate conduction in the valves, with the result that a converter in a control or regulating circuit functions as an amplifier. A somewhat greater amount of energy is necessary to reinstate the blocking condition of the valves—i.e. to interrupt the current. When an external a.c. source is used to turn off the valves the converter is said to be naturally commutating. According to the nature of this a.c. source, a further subdivision may be made into supply-commutating and load-commutating converters. In self-commutating converter circuits the valves are turned off by internal commutation circuits.

Most converter circuits embody a number of valves, which take over the flow of current periodically in the course of the operating cycle. The transfer of current from one valve to another is called 'commutation'. Commutation is one of the most important concepts in converter circuit theory. Since the externally commutating converter requires no special commutation circuit it can be considered as having natural commutation. The self-commutating circuits, on the other hand, are known as converters with forced commutation.

THE STRUCTURE OF CONVERTERS

The most important components of converters are the valves. By grouping together a number of valves, any desired power rating may be achieved. In converter assemblies the valves are mounted on isolating cooling structures and provided with associated control and protection devices.

A converter equipment includes in addition the control electronics, the matching transformer, the smoothing and filter circuits, and, in the case of self-commutating systems, the forced-commutation circuits.

APPLICATION AREAS OF CONVERTERS

In industry, converters are used principally as controllers for electric motor drives and as d.c. sources for electrolytic processes.

In railways and tramways, converters are used both in substations, as d.c. sources, and in vehicles, as controllers.

In power generation and transmission, converters serve as rectifiers for generator excitation or as a.c.–a.c. converters for generator starting, network links, reactive power compensation, or high-voltage d.c. transmission.

THE PRINCIPAL FEATURES OF POWER ELECTRONICS

A great many different kinds of converter of different power ratings can be provided by relatively few types of valve. This facilitates a high degree of flexibility and adaptability in the designing of equipment and a notable simplification in the operating of plant.

The active element of the converter valve is made from a thin silicon wafer which handles the whole of the converter power. The power density in these valves is very high—the volume of the semiconductor material is from four to six orders less than that of the complete converter unit. This extreme concentration of power in turn makes extreme demands upon the cooling and protection of the valves.

The operation of converters will be explained in terms of transient electrical processes.

The control and conversion of electrical energy is achieved with high efficiency, without inertia, and without moving parts

PROBLEM AREAS OF POWER ELECTRONICS

Power electronics can be categorized in terms of problems and areas of interest as follows:

Converter valves	— physics
	— manufacture
	— testing and measurement
Converter modules	— construction
	— cooling
Converter circuits	— design
	— circuit theory
	— system characteristics
Converter protection	— overcurrent protection
	— overvoltage protection
Control and regulation	— pulse generation (gate triggering)
	— pulse transmission
	— control circuits
Supply loading	— harmonics
	— reactive power
Converter installations	— process control
	— filters

2

The History of Power Electronics Development

MERCURY ARC RECTIFIERS

At the beginning of the twentieth century, in the year 1902, P.C. Hewitt (USA) discovered the rectifying effect of a mercury vapour arc and constructed the first mercury arc rectifier. The negative electrode, the cathode, was in the form of liquid mercury in the lower part of the evacuated glass tube. The positive electrode, the anode, fitted into the upper part of the tube, was of pure graphite.

By means of an igniting device, an electron-emitting area, the cathode spot, was formed on the surface of the mercury and the mercury vapour was thereby ionized. Since the ionized vapour, the arc plasma, permitted the passage of charge carriers only in one direction, the arrangement behaved as a converter valve.

Almost simultaneously with the invention of the mercury arc rectifier, in 1904, the vacuum diode was developed by J.A. Fleming. The vacuum diode, however, could rectify only very small currents, in contrast to the mercury arc valve which had almost reached the hundred-ampere limit.

In 1909, B. Schaefer (Germany) constructed the first steel-tank mercury arc rectifier in the shape of a single-anode water-cooled tube, furnished with a vacuum pump. By this means it was possible to increase the current loading of mercury arc rectifiers to several hundred amperes.

Industrial production of steel mercury arc rectifiers began before the First World War. These were six-anode water-cooled tubes with a large mercury vapour condensation chamber, capable of handling up to a hundred kilowatts.

The first mercury arc rectifiers were uncontrolled. The control facility, by virtue of a gate incorporated between the anode and the cathode, was introduced by Langmuir (USA) in 1914. The similarly constructed vacuum triode had already been invented by de Forest in 1906.

The first phase of development of the mercury arc rectifier reached a climax in the mid-thirties with the production of multi-anode (up to 24 anodes per tube) water-cooled 'giant' tubes for currents of up to 8000 A, weighing up to 15 000 kg. At the same time, however, there began the development of the single-anode mercury arc valves known as the ignitron (by J. Slepian, USA) and the excitron, and also of pumpless tubes (Daellenbach and Gerecke, Switzerland).

As well as the term 'rectifier', the term 'converter' (*Stromrichter* or *Mutator*) was current in German-speaking areas since the early thirties.

The most important applications of mercury arc converters up to the time of the Second World War were to electric tramways and railways and to electrolytic plant.

After the Second World War, production was mainly of single-anode and six-anode mercury arc converters in pumpless form. In addition to electrolytic processes and d.c. traction, mercury arc converters were used from the mid-fifties in a.c. locomotives (in the form of ignitrons) and in d.c. motor drives, where they replaced rotary converters.

In the early sixties, mercury arc converters were replaced by semiconductor converters, first in electrolytic equipment and traction and subsequently in motor drives.

Although the mercury arc converters had developed, in their fifty-odd years of existence, into robust and reliable power electronic elements, they were unable, like the vacuum tubes in the telecommunications field, to compete with the new semiconductor technology. The last application of the mercury arc converter was to high-voltage d.c. transmission in the nineteen-sixties. Vacuum tubes, on the other hand, are still manufactured for special purposes (transmitter tubes, cathode-ray tubes, etc.), as are high-power thyratron switches.

SEMICONDUCTOR VALVES

That the semiconductor–metal junction possessed rectifying properties was observed by Braun (Germany) and Schuster (Great Britain) as early as 1874, but for a long time it found no practical application. (The term 'semiconductor' was not used by Braun and Schuster, but had been used by Volta in 1783.) The point-contact crystal detector was first used at the beginning of the twentieth century in the then-developing wireless communications engineering. The first semiconductor power rectifiers, however, were produced only in the early thirties, based on a metal–selenium interface. These 'dry-plate' rectifiers could at that time compete with the mercury arc rectifier only in the lowest range of voltages.

Germanium and silicon point-contact diodes came into being in the early forties (for radar equipment). In 1948 the transistor effect in the germanium crystal was discovered, and a year later the theory of the p–n junction was perfected.

Germanium power diodes date from 1954. Three years after that the development of semiconductor technology in the high-power field reached its culmination—the controllable silicon valve, the thyristor, was developed. So began the true age of power electronics. In the early nineteen-eighties practical gate turn-off power thyristors were developed. These are mainly applied to converters with forced commutation.

Although mercury arc rectifiers made remarkable progress during the nineteen-sixties (1 kA/10 kV per valve), within a decade they were completely superseded by semiconductor valves, whose parameters were much more modest (0.2 kA/2 kV per valve). The reasons lay in the higher efficiency, smaller dimensions, greater convenience, and greater reliability in operation of the semiconductor valve.

In construction the silicon valve developed from compact directly cooled elements, through stud-based and flat-based cells, to the capsule form with separate double coolers. The crystal diameter meanwhile grew sixfold from the original 20 mm.

In contrast to the mercury arc valves, whose development progressed quite empirically, in the case of semiconductor valves technological success was underpinned by comprehensive theory.

CONVERTER CIRCUITS AND APPLICATION AREAS

The first rectifier circuit, the single-phase bridge, originated with Pollak and Graetz (Germany) in 1896–1897. The valves used were electrolytic cells.

The theoretical foundations of converter equipment were laid down by Daellenbach and Gerecke (Switzerland) in 1924. Practically all the most important converter circuits, such as the direct frequency changer, the d.c.–link converter, the forced-commutation inverter, etc., had been evolved by the beginning of the thirties.

Mercury arc rectifiers played a dominant role in the application of power rectifiers until the mid-fifties. They were installed principally in railway and tramway substations, rectifier locomotives, and electrolytic plant. In their last decade of production mercury arc converters were installed increasingly in d.c. motor drives in place of rotary converters (hence: *static converters*). The control devices used in mercury arc converters were magnetic amplifiers (transductors).

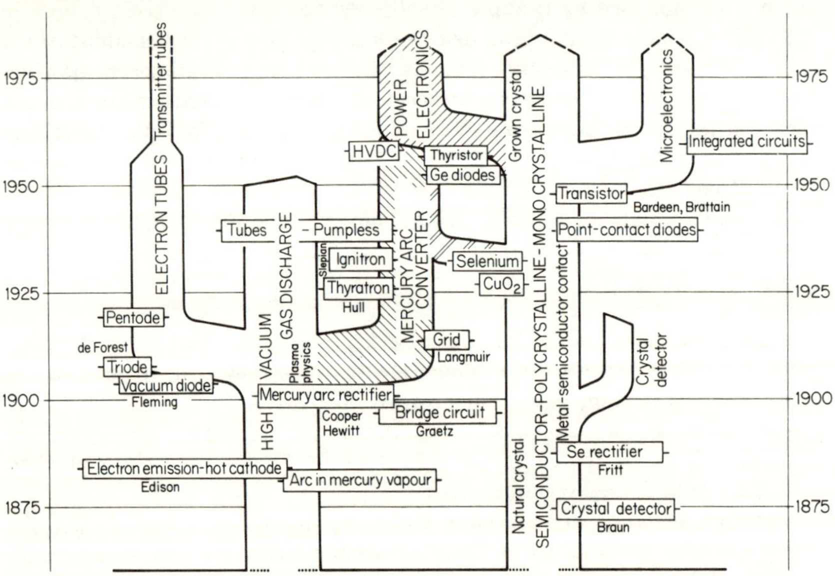

Figure 3. Diagrammatic presentation of the history of power electronics development. In the first period (1910–1960), the gas-discharge valve (mercury arc converter) was predominant. In the second development period (since ca. 1960), semiconductor valves (thyristors and diodes) have played the major role

Following the invention of the thyristor, there began in the mid-sixties a rapid development in converter technology. Power electronics spread into nearly every field of electrical engineering: in industry, in transport, in electrochemistry, and in power transmission.

Unlike mercury arc valves, which were manufactured individually, thyristors (and of course diodes) were mass-produced. The advent of semiconductor control and regulation techniques now made practical nearly all the converter systems that had previously found no application, and made it possible to choose the optimum circuits. (The multi-anode type of mercury arc converter, with a common cathode, influenced circuit design. Use was made principally of single-way circuits.)

Figure 3 presents a summary of the development history of power electronics, embracing also the parallel developments in the field of telecommunications.

3

Converter Valves

FUNDAMENTALS

Converter valves are electronic components that pass current in only one direction.

Converter valves have only two stable operational states: they are either blocking or conducting.

In circuit theory converter valves are treated as ideal switches.

PHYSICAL PRINCIPLES

Valve action can be produced in various physical media:

1. In air, by means of vibrating contacts—the contact rectifier
2. In vacuum, through electron emission from a hot cathode–the vacuum diode
3. In a gas, through a stream of ions—the mercury arc valve, the thyratron, the ignitron
4. In a liquid, through a stream of ions—the electrolytic cell
5. At an interface between a semiconductor and a metal, by a blocking layer effect—selenium and copper oxide rectifiers
6. In a single-crystal semiconductor, through charge carriers in layers of different conductivity types—silicon diodes and thyristors

For two decades, for practical purposes only silicon devices have been used in power electronics.

SEMICONDUCTOR VALVES

There are principally two types of valve in use in power electronic circuits with natural commutation:

1. Diodes, uncontrolled semiconductor valves with two electrodes—anode and cathode
2. Thyristors, controlled semiconductor valves with three electrodes—anode, cathode, and gate

With reverse voltage applied the two valves behave in the same way; they do not conduct. With forward voltage applied the diode always conducts; the thyristor, on the other hand, conducts only when a triggering pulse has been applied to it between the gate and the cathode. Once a valve has been switched on and is carrying current it can no longer be controlled by the gate. The turning-off and blocking of valves—diodes or thyristors—is thus dependent only upon the characteristics and condition of the anode circuit. The interruption of current is brought about by external means.

CONSTRUCTION OF SEMICONDUCTOR VALVES

Extremely pure silicon is used in the manufacture of power diodes and thyristors. Silicon is one of the group of materials classified as semiconductors, whose conductivity lies between that of conductors and insulators. In contrast to conductors, whose resistance increases with increasing temperature, in semiconductors an increase in temperature causes an increase in conductivity. In semiconductors the charge carriers necessary for the conduction of current become free only as a result of an input of energy.

The silicon material is produced in monocrystalline form with a completely symmetrical internal structure. By means of a controlled addition of impurity atoms, two different conduction types are obtained:

1. Silicon material with an excess of negative charge carriers—electrons (abbreviated to 'n-type')
2. Silicon material with an excess of positive charge carriers, which result from electron deficiencies, or holes (abbreviated to 'p-type')

A silicon wafer with a two-layer p–n structure exhibits valve characteristics. It constitutes the active element of a diode.

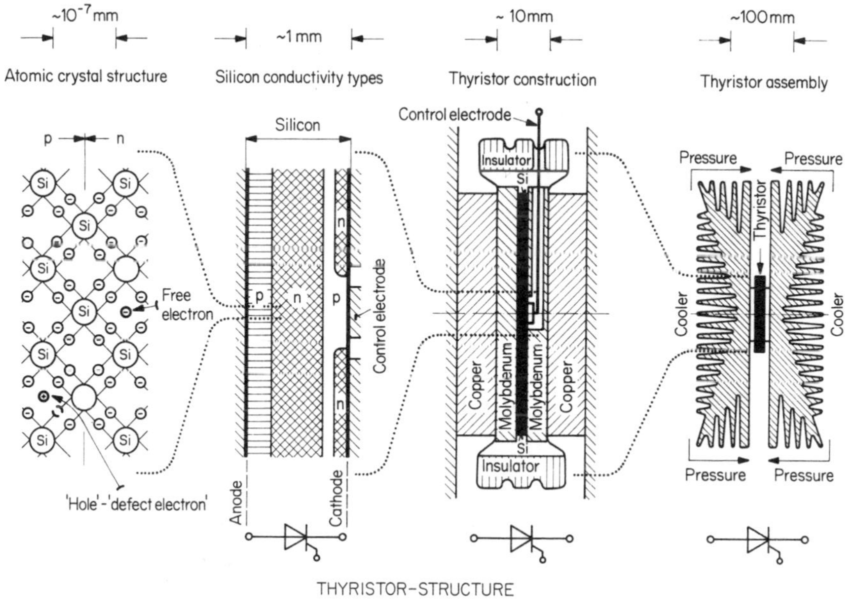

Figure 4. The construction of the controlled converter valve (thyristor) shown to four different scales—from the thyristor-cooler assembly to the atomic structure. The function of the valve takes place in the thin silicon wafer with four conductivity layers (p–n–p–n)

A four-layer p–n–p–n structure functions as a controllable valve—a thyristor. The outer p-layer is designated the anode, the outer n-layer the cathode.

The silicon wafer, the active part of the converter valve, is mounted between two discs of molybdenum, which has a similar coefficient of thermal expansion and high conductivity and is contained in a mechanically robust, insulated, and gas-tight cell. The silicon is very hard, but brittle. The cell complete with silicon wafer constitutes the basic component of power electronics—the semiconductor device. The device is clamped between the two halves of a cooler. Very high clamping pressure ensures good electrical and thermal contact.

Figure 4 illustrates the construction of a thyristor diagrammatically at four levels, from the crystal structure to the cooler assembly.

OPERATION OF DIODES AND THYRISTORS

The diode

The diode is made from a silicon crystal wafer with layers of n– and p–type conduction characteristics. The two layers form an interface within the crystal, either directly or, in the case of power diodes, through a high-resistance intermediate layer. When a negative voltage is applied between the anode (p–layer) and cathode (n–layer), the p–n junction is depleted of charge carriers; the charge carriers are attracted to the opposite polarities of the electrodes. The diode blocks, the voltage is supported by the p–n junction, and no current flows. If the voltage source is reversed, so that a positive voltage is applied between the anode and cathode, the p–n junction and the two layers are flooded with charge carriers (electrons and 'holes'). The diode conducts and current can flow.

As long as current is supplied by the external circuit, the diode continues to conduct. If the current falls to zero and the voltage between anode and cathode becomes negative, the valve blocks. No current flows in the negative direction.

The thyristor

The thyristor is made with four layers, p–n–p–n; i.e. with three p–n junctions. When a negative voltage is applied between the anode and cathode, the two outer p–n junctions are depleted of charge carriers and support the voltage, so that no current can flow. With a positive applied voltage between the anode and cathode, the middle junction is depleted and prevents the flow of current. This blocking effect, however, can be removed by a triggering pulse applied to the inner p-layer (the gate). The triggering pulse causes a sudden flooding of the whole crystal with charge carriers ('firing') and the thyristor switches from the blocking to the conducting state. In the conducting state, as in the negative blocking state, the thyristor behaves exactly like the diode.

CHARACTERISTICS OF DIODES AND THYRISTORS

The electrical properties of the diode and the thyristor can best be explained in terms of their current–voltage characteristics.

Figure 5 shows the three most important representations of diode and thyristor characteristics:

The ideal characteristics—the valve is assumed to operate as a switch
The static characteristics—corresponding to d.c. measurements
The dynamic characteristics—representing actual operation

The full switching cycle of the thyristor can be understood from Figure 6. The individual switching phases, or operating conditions, are illustrated by the characteristics corresponding to the waveforms. The structure diagrams indicate the depletion or suffusion of the p–n junction with charge carriers. Also shown are the corresponding equivalent circuits of the thyristor—actual and ideal. It will be seen that a different equivalent circuit is associated with each switching stage.

For the purposes of converter circuit theory, ideal valve characteristics are quite adequate. For switching problems, however, it is necessary to take into consideration the dynamic characteristics and the actual equivalent circuits.

Once the thyristor is conducting, the current cannot be switched off or influenced by gate current. The current zero is brought about by the circuit.

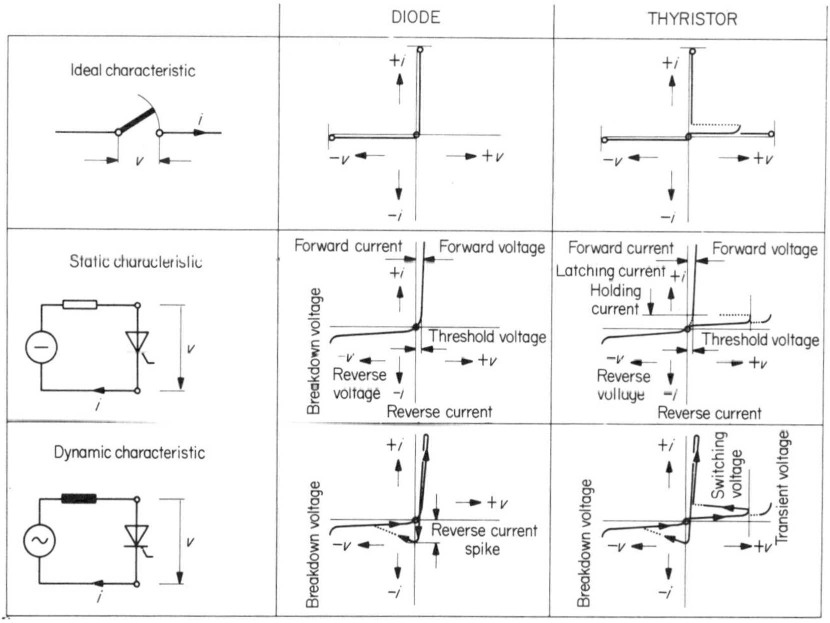

Figure 5. Characteristics of diodes and thyristors. Ideal characteristics—valves regarded as switches—for circuit theory. Static characteristics—d.c. measurements—for data sheets. Dynamic characteristics—typical behaviour of valves in operation in a converter circuit

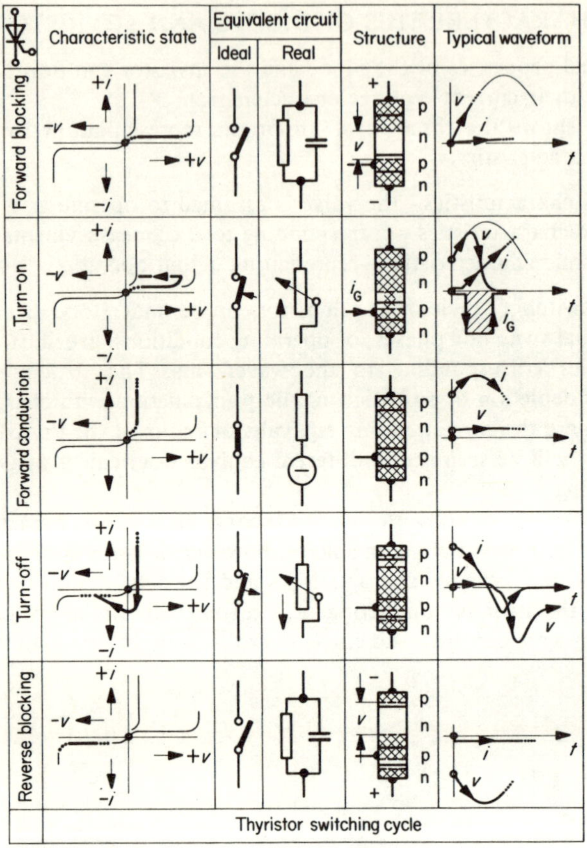

Figure 6. The complete operating cycle of the thyristor from blocking forward (positive) voltage, through switching-on by a triggering pulse and consequent conduction of forward current, to turning-off and blocking reverse (negative) voltage

THE THYRISTOR COMPARED WITH THE TRANSISTOR

The transistor is the most important component in electronic engineering. It is made with three layers, p–n–p (or n–p–n), and thus stands from a structural point of view between the two-layer diode and the four-layer thyristor. The transistor, however, cannot be considered as a converter valve, since it can support voltage only of one polarity. In contrast to the situation in the thyristor, whose current in the conducting state cannot be controlled through the gate, the main (collector) current of the transistor is always under the influence of the control (base) current. The structure, characteristics, and operation of the transistor in comparison to the thyristor and the diode is illustrated in Figure 7.

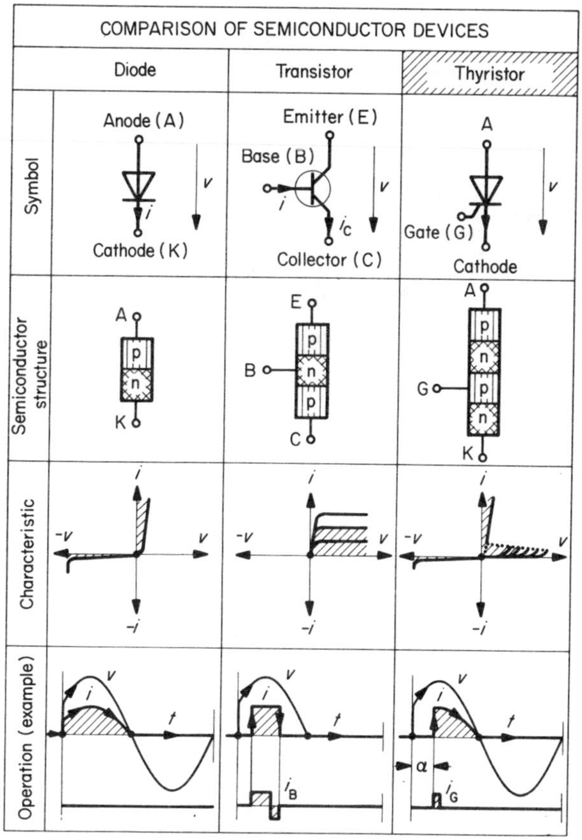

Figure 7. Comparison of the internal structure, characteristics, and mode of operation of uncontrolled and controlled converter valves—diodes and thyristors—with those of the transistor. In the transistor the current can be regulated continuously by the control signal, in contrast to the thyristor, in which only the instant of turning-on can be controlled

OPERATIONAL VALVES

The diode is in itself basically operational, in that it will conduct current or block. Nevertheless, it must in use be:

1. Mounted, connected electrically, and insulated
2. Cooled (by air or liquid)
3. Protected against overcurrent and overvoltage

The thyristor is by itself not basically operational. Without triggering pulses it blocks continuously with either polarity of voltage. As well as the additional requirements mentioned above for diodes, the thyristor needs a pulse generator to produce periodical repetitive pulses, a firing unit to shift the

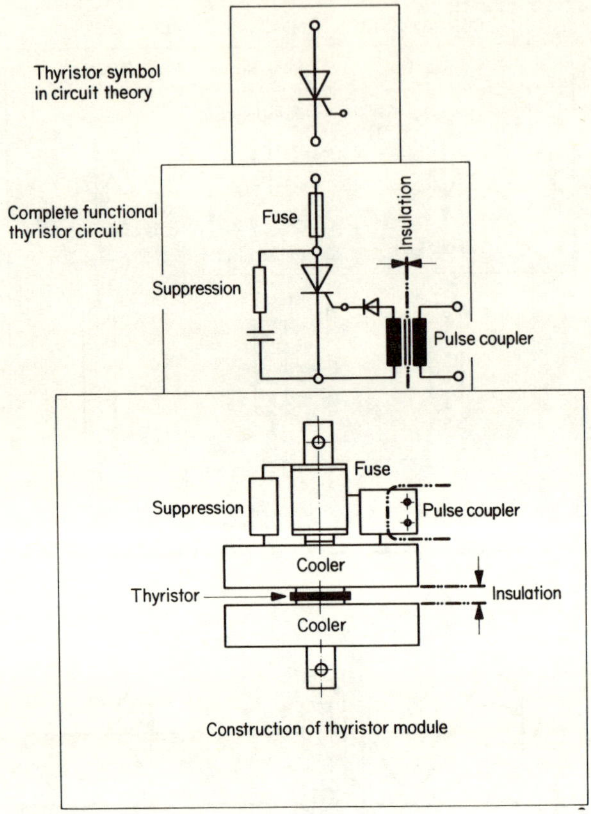

Figure 8. A simple thyristor symbol (top) in converter circuit theory actually represents a complex functional unit. In the study of converter circuits the protection devices and the means of control are usually omitted for the sake of simplicity

pulses in phase (the two assembled as one control unit), and a pulse coupler (mounted close to the thyristor) to isolate the control circuit from the power circuit.

Figure 8 shows as an example the circuit and the construction of an operational thyristor unit. It will be observed that the necessary cooling, protection, and control components are in total much larger than the actual valve.

VALVE GROUPS

Although the current and voltage loading of individual diodes and thyristors are strictly limited in accordance with their ratings, a valve group can handle practically unlimited power.

A *parallel* connection of the valve increases the current capability and a *series* connection increases the voltage capability.

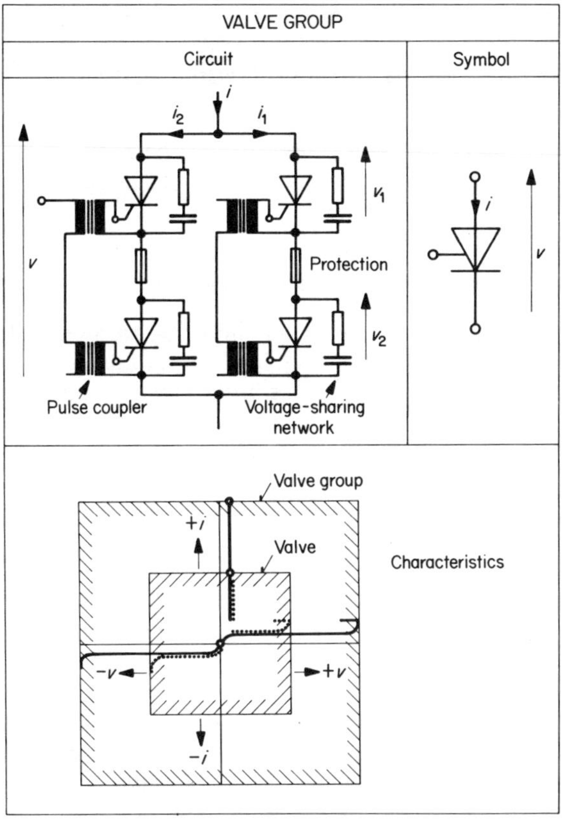

Figure 9. Parallel and series-connected thyristors (or diodes) form a valve group with current and voltage ratings that are multiples of those of the single valve. For the sake of simplicity the valve group is represented by a single symbol in converter circuit theory. Example with two series pairs of valves in parallel

In a converter circuit a valve group behaves like an individual valve. In the group it is necessary to ensure:

1. *Current sharing* between parallel-connected valves
2. *Voltage sharing* between series-connected valves
3. *Simultaneous triggering* of all valves by the control pulses

As an example, Figure 9 shows a valve group with two parallel-connected and two series-connected thyristors.

VALVE PARAMETERS

The principal valve parameters are:

Permissible blocking voltage (maximum permissible instantaneous value)

Permissible forward current (mean value, depending upon waveform and cooling)
Switching power (product of forward current and blocking voltage: little used)

A valve contributes a certain proportion of the power transmitted by the whole converter. This proportion is normally much less than the 'switching power'.

DYNAMIC CHARACTERISTICS OF CONVERTER VALVES

In contrast to an ideal switch with infinitely fast turn-on and turn-off, the turning-on and turning-off processes of real valves take place at a finite rate.

Diodes and thyristors do not block instantly when the forward current reaches zero, but conduct for a short time in the reverse direction (the *carrier storage effect*). They turn off only after a short delay period. A relatively high reverse current spike builds up and turn-off does not, therfore, take place at zero current. The reverse current spike depends upon the rate at which the forward current falls. The rapid cessation of reverse current gives rise to overvoltages in the circuit inductance, which endanger the blocking valve. The converter valve fundamentally exhibits a tendency to self-destruction.

After the current zero, a certain time elapses before the valve blocks (the *reverse recovery time*). In the case of thyristors, a further period elapses before they will block forward voltage (the *forward recovery time*).

When a thyristor is turned on by a gate pulse the current spreads across the area of the silicon wafer with finite velocity. The *rate of rise of current* (di/dt) is thereby limited because very high rates of rise can result in local overstressing of the silicon wafer.

A very high *rate of rise of voltage* in the forward blocking state (dv/dt) can cause the thyristor to turn on without a gate pulse, with a consequent risk of failure.

In the zero-current condition and in the reverse or forward blocking state valves behave like capacitors.

TURN-ON REQUIREMENTS OF THYRISTORS

The thyristor will turn on when, with a positive applied voltage which must exceed a certain minimum value, a positive current pulse of sufficient amplitude, and rate of rise of at least a certain minimum duration, is applied to the gate. Within the duration of the pulse, the forward current must attain a certain minimum level (latching current) to ensure that the thyristor will remain conducting.

TEMPERATURE DEPENDENCE OF VALVE CHARACTERISTICS

The physical processes in the silicon crystal are temperature dependent. Hence all the properties and characteristics of a valve depend upon the temperature of the silicon crystal. With increasing temperature, the leakage current increases, and in a thyristor the forward-blocking capability decreases. Normally the maximum operating temperature of a diode is about 145 °C and that of a thyristor about 125 °C. Much higher temperatures are permissible for short periods, however, when no voltage is applied to the valves.

LOSSES IN VALVES

Valves are heated in operation as a result of losses in the active element—the silicon wafer. In the course of the switching cycle, dissipation occurs in the valve from the following sources:

1. *Turn-on losses* (in thyristors)
2. *Conduction losses*
3. *Turn-off losses*
4. *Blocking losses*

In the case of thyristors operating from the supply mains (up to 60 Hz), the conduction losses produced by forward current are the largest. Although the losses in the valve are very small in comparison with the power throughput, the specific loading of the active valve elements is very high because the volume and the surface area of the disc are very small.

LIMITS OF SURVIVAL, OVERLOAD CAPACITY, AND RELIABILITY OF VALVES

The silicon wafer is very thin; the internal electric field strength in the blocking condition of the valve is therefore very high. When the reverse voltage exceeds a certain critical value, avalanche breakdown occurs, with a large increase in current, and if sufficient energy is available from the voltage source this will lead to a local dielectric breakdown and puncture of the silicon wafer. This represents a catastrophic failure of the valve and current then flows in both directions, as in a conductor. A forward-blocking voltage in excess of the breakover level can also result in the destruction of the thyristor. Following the turn-on the active element may be destroyed by the high di/dt. A large forward overcurrent can also destroy the valve through excessive temperature.

The volume of the active semiconductor material in a valve is very small and therefore has very little thermal capacity. This has a great effect on its overcurrent capability, particularly for short periods. For longer periods it is the thermal capacity of the cooler that is significant.

So long as the safe operating limits of current, voltage, and temperature are observed, semiconductor converter valves are very reliable. Cyclic loading is, however, an unfavourable condition compared to constant load.

STATISTICAL DISTRIBUTION OF VALVE PARAMETERS

Because of the complexity of manufacture of semiconductor valves, the parameters and characteristics of diodes and thyristors, of any particular type, exhibit a relatively large spread. The data and characteristics provided by the manufacturer accordingly represent typical values, examples, or limits. For special purposes and applications (series or parallel operation) valves are often classified further into subgroups.

SPECIAL VALVES

Besides the conventional diodes and thyristors, which are intended mainly for use in supply-commutating converters, there are various valves for special applications:

High-frequency thyristors (fast-switching thyristors)—with low switching losses and low turn-off time; intended for higher operating frequencies and forced commutation

Reverse-conducting thyristors—behave in the same way as a conventional reverse-blocking thyristor with an inverse parallel diode, e.g. in d.c. switching regulators (choppers)

Bidirectional thyristors (triacs)—behave like two conventional thyristors in inverse parallel

Gate turn-off thyristors (GTO)—can be turned off by a negative gate current

4

Simple Valve Circuits

PRINCIPAL ELEMENTS OF CONVERTER CIRCUITS

In the theory of supply- and load-commutating converters it is usual to make the following assumptions and simplifications:

1. The valves are ideal.
2. The supply mains, transformers, and machines are lossless and without capacitance.
3. A.C. and d.c. voltage sources are ideal.
4. Inductances are linear.

For the purposes of a theoretical investigation, therefore, converter circuits are mostly composed only of the following elements:

1. Valves (thyristors or diodes)
2. Inductances
3. Voltage sources (a.c. or d.c.)

In order to understand the fundamental operation of converters, it is useful first to examine the behaviour of simple valve circuits.

THE TRANSIENT STATE—CONTINUOUS OCCURRENCE IN VALVE CIRCUITS

In an a.c. circuit with inductance and resistance, the current under steady-state conditions is ideally sinusoidal and lags the voltage by an electrical angle ϕ. This condition, however, is only established some time after the circuit is switched onto the supply. In the period immediately after switching on, the current waveform generally departs considerably from the ideal symmetrical sinusoid.

This is illustrated in Figure 10 (top). Figure 10 (bottom) shows how conditions are changed when an uncontrolled valve—a diode—replaces the switch. When the switch is closed (it is assumed that, like the diode, the switch closes at a voltage zero) the current waveform is initially the same. In the diode circuit, however, current cannot flow in the negative direction and the circuit is therefore broken at the current zero. In the second cycle the process starts again from the beginning and the current waveform in the second and all subsequent cycles is similar to that of the first, in contrast to the circuit with the switch, in which the current waveform changes with time.

The valve circuit is thus continuously in the transient state.

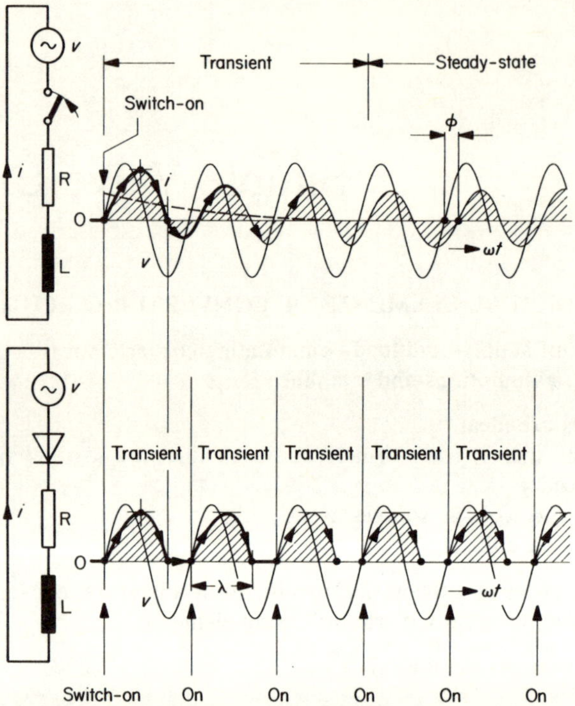

Figure 10. Comparison of the behaviour of a valve circuit (bottom) with that of a normal a.c. circuit. After switching on, the current in the a.c. circuit settles, in time, into the steady-state condition. The valve circuit remains permanently in the transient condition

INDUCTIVE VALVE CIRCUIT WITH A.C. SUPPLY

In an a.c. circuit in the steady state the current i in the inductance L lags the voltage across the inductance v_L by approximately 90°el. The current reaches its maximum value at the voltage zeros and the maximum voltage occurs at the current zeros. This is in accordance with the general law of inductance: the voltage v_L across the inductance is proportional to the rate of change of current di/dt and the inductance L.

When the rate of change of current di/dt is zero, as it is at the current peak, the voltage v_L is also zero. The greatest rate of change of current is at the current zero, and at that point the voltage is at a maximum. If the voltage is considered to be sine function, the current is of cosine waveform. The requirements of the inductance law quoted above must also be met in the transient condition after switching on. In addition, the current must be equal to zero at the instant of switching on.

All these conditions are met by a current waveform that is phase-shifted 90° with respect to the voltage and displaced vertically so that it increases positively from zero at switch-on. This is shown in Figure 11. Switching on

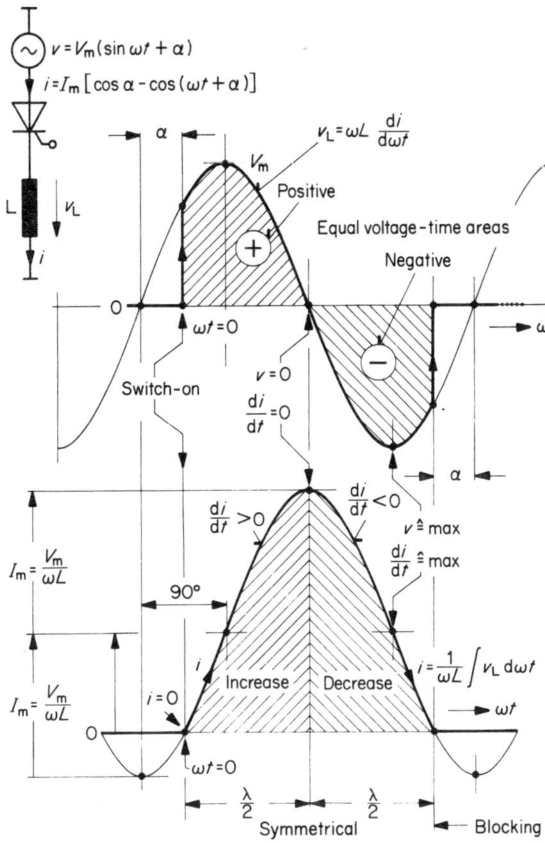

Figure 11. Ideal inductive valve circuit with a.c. voltage source. After the thyristor has been triggered at angle α the full sinusoidal source voltage appears across the inductance L (top). The current waveform is also sinusoidal, but displaced in time (by 90° el) and vertically. α is termed the delay angle

occurs in the diagram at an electrical angle α (angle of delay) measured from the first voltage zero. The gate of the thyristor receives a triggering pulse at this point. (The control circuit is omitted for simplicity.)

Since an ideal circuit is assumed, the voltage across the inductance after switching on is equal to the supply voltage.

The current waveform is in the form of a cosine wave symmetrically disposed about the second voltage zero. During the period of positive voltage the current builds up to its maximum, and in the equal period of negative voltage it is reduced to zero.

Current cannot flow in the negative direction because of the valve effect in the thyristor, so the process is then interrupted. The next cycle proceeds in exactly the same way (assuming an unchanged delay angle α).

Figure 12 shows all the voltages and currents in the inductive valve circuit

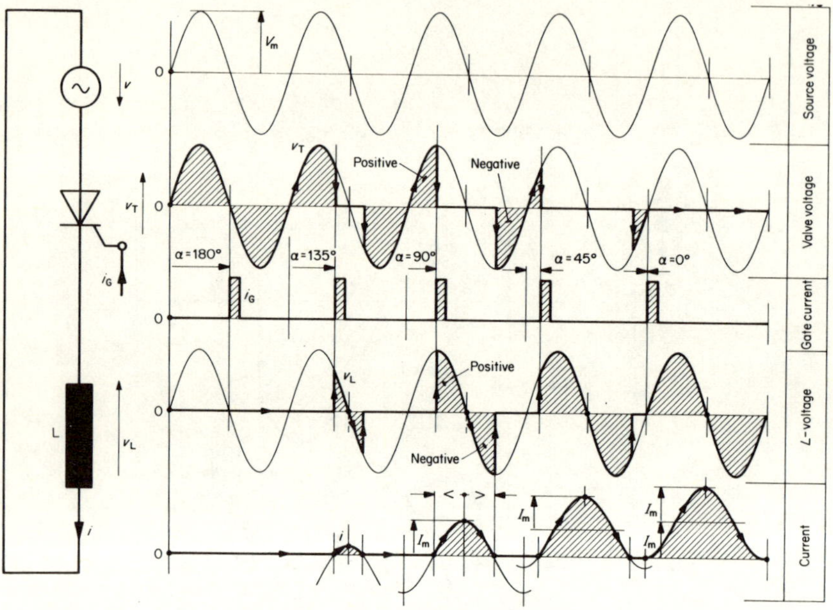

Figure 12. Operation of a valve circuit with a.c. voltage source, thyristor, and inductance, with variable delay angle α. The current flows only in one direction: it is rectified. The inductor voltage v_L, on the other hand, is an alternating voltage, with equal positive and negative voltage–time integrals

for the whole control range ($0 < \alpha < 180°$). It will be observed that although the voltages are alternating (with a mean value of zero) the current is unidirectional. The circuit behaves as a *rectifier*. The current flows in the positive direction even when the applied voltage is negative (the stored magnetic energy in the inductance being released). Formulae for calculating the mean and r.m.s. currents—together with expressions for the resistive valve circuit—are given in Figure 13.

MATHEMATICAL APPENDIX I

The source voltage is defined as

$$v = V_m \sin(\omega t + \alpha) \tag{1}$$

where V_m = amplitude of sine wave
ωt = electrical angle
$\omega = 2\pi f$ = angular frequency
f = frequency
α = delay angle (zero on the ωt axis)

The voltage across the inductance is given by

$$v = L\frac{di}{dt} = \omega L \cdot \frac{di}{d\omega t} \tag{2}$$

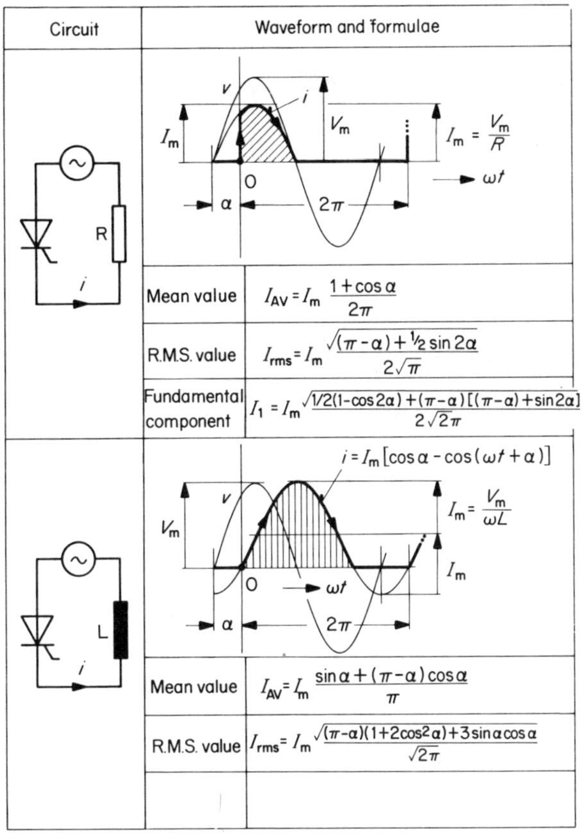

Figure 13. Summary of the principal formulae relative to the inductive and resistive valve circuits with an a.c. voltage source. With resistive load (top) the current equations are similar to those for the resistor voltage

Here di/dt is the rate of change of the current i in the inductance. After switching on, in the inductive valve circuit (Figure 11)

$$v_L = v$$

Rearranging the above equations,

$$\omega L \frac{di}{d\omega t} = V_m \sin(\omega t + \alpha)$$

and the current is then

$$i = \frac{V_m}{\omega L} \int \sin(\omega t + \alpha)\, d\omega t \geqslant 0$$

Integration gives

$$i = I_m [\cos \alpha - \cos(\omega t + \alpha)] \qquad (3)$$

$$I_m = \frac{V_m}{\omega L}, \qquad i \geqslant 0 \tag{4}$$

For $\alpha = 0^\circ$ (thyristor fully conducting, or diode), this becomes

$$i = I_m(1 - \cos \omega t) \tag{5}$$

The mean value of the current is calculated as follows:

$$I_{AV} = \frac{1}{2\pi} \int_0^{2(\pi - \alpha)} i \, d\omega t \tag{6}$$

Substitution and integration give

$$I_{AV} = \frac{I_m}{\pi} [\sin \alpha + (\pi - \alpha) \cos \alpha] \tag{7}$$

The r.m.s. current is calculated in accordance with the definition

$$I_{rms} = \sqrt{\frac{1}{2\pi} \int_0^{2(\pi - \alpha)} i^2 \, d\omega t} \tag{8}$$

which gives

$$I_{rms} = \frac{I_m}{\sqrt{2\pi}} \sqrt{(\pi - \alpha)(1 + 2\cos^2\alpha) + 3 \sin \alpha \cos \alpha} \tag{9}$$

With the delay angle $\alpha = 0^\circ$ this becomes

$$I_{rms0} = I_m \sqrt{\frac{3}{2}} \tag{10}$$

and

$$I_{AV0} = I_m \tag{11}$$

INDUCTIVE VALVE CIRCUIT WITH A.C. AND D.C. VOLTAGE SOURCES

If an inductance L is connected to a d.c. source of voltage E through a diode (Figure 14, top), the current i increases linearly with time (resistance is neglected). The higher the voltage and the smaller the inductance, the more steeply the current rises.

In the inductive a.c. circuit including a diode (Figure 14, bottom), the current after switching on follows a cosine waveform with a displaced zero.

In all subsequent cycles the current waveform is repeated, so that, in contrast to Figure 14 (top), the current level does not increase.

In a circuit without a valve, under steady-state conditions, the current has the same waveform—a sine wave phase-shifted 90° with respect to the voltage—as that shown in Figure 14 (bottom), but without the vertical displacement of the zero axis.

If both voltage sources are now connected to one valve circuit, so arranged that the d.c. source opposes the flow of current in the valve, then the waveform of the current i is as illustrated in Figure 15. The resultant current is the difference between the two hypothetical currents which would be produced by the two voltage sources acting independently: $i = i_{\sim} + i_E$.

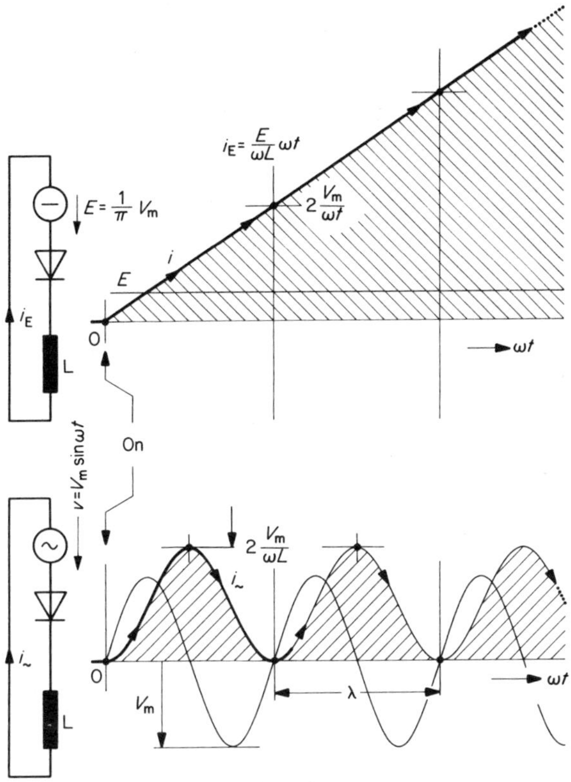

Figure 14. Operation of the inductive valve circuit with a d.c. voltage source (top) and an a.c. voltage source (bottom). The valves are assumed to be diodes. The direct voltage (E) equates with the mean value of the rectified alternating voltage (V_m/π)

However, in contrast to Figure 14, the current starts to flow only when the voltage v becomes more positive than the voltage E. During the time interval λ the difference between the two voltages is applied to the inductance: $v_L = v - E$. The current maximum occurs at the point of intersection $v_L = E$. To reduce the current to zero requires the same voltage–time area applied to the inductance as that which causes the increase in current; the period of negative voltage is not, however, of the same duration as that of the positive voltage (the voltage falls more steeply) and the current takes less time to fall than to rise.

The conduction period λ depends upon the ratio E/V_m. The higher the d.c. voltage, the shorter is the conduction period. When $E = V_m$ no current flows at all.

If the d.c. voltage is reversed, the current is the sum of the two components $i_\sim$ and i_E; it therefore (with an uncontrolled valve) never falls to zero, but increases continuously (Figure 16, top).

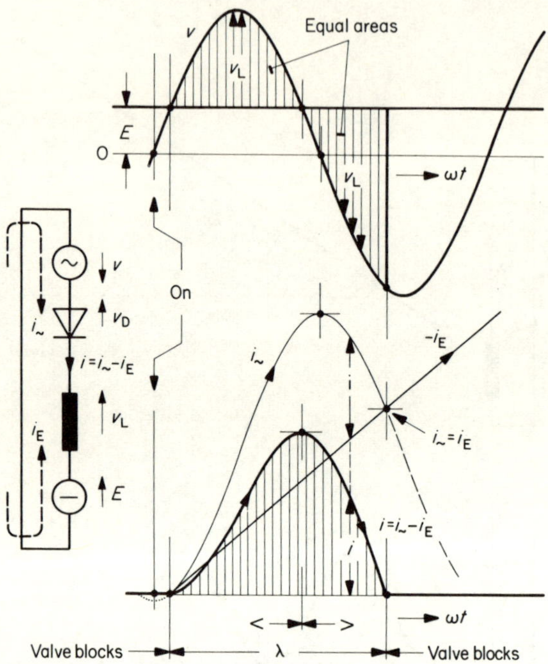

Figure 15. Inductive valve circuit with two (d.c. and a.c.) voltage sources. The current waveform can be constructed as the difference between the two current components ($i_\sim$ and i_E) supplied by the two sources v and E (cf. Figure 14)

MATHEMATICAL APPENDIX II

From the inductance law, the voltage v across the inductance L is proportional to the rate of change of current di/dt:

$$v_L = \omega L \frac{di}{d\omega t}$$

The current is therefore given by

$$i = \frac{1}{\omega L} \int v_L \, d\omega t \tag{12}$$

After switching on, the voltage applied to the inductance in Figure 14 (top) is constant:

$$v_L = E$$

By substitution in equation (12)

$$i_E = \frac{E}{\omega L} \omega t \tag{13}$$

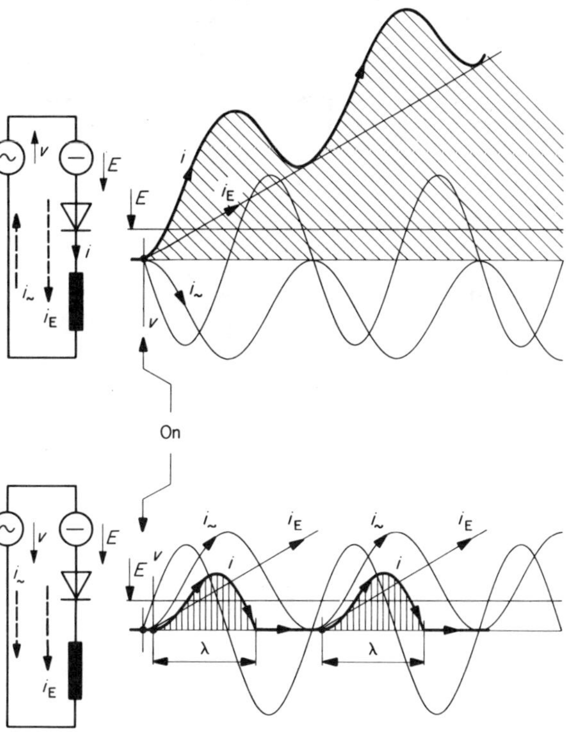

Figure 16. Inductive valve circuit with a.c. and d.c. voltage sources of different polarities. The current components $i_\sim$ and i_E supplied by the two sources may either add (top) or, as in Figure 15, subtract, depending upon the polarity

CURRENT INTERRUPTION IN AN INDUCTIVE VALVE CIRCUIT—COMMUTATION AND FREEWHEELING

The converter valve, whether it is uncontrolled (diode) or controlled (thyristor), remains conducting so long as the external circuit drives current through it in the forward direction. The turn-on instant of a thyristor can be determined easily and with very little power by a triggering pulse, but turning off is effected by the circuit and can therefore be influenced only by varying the circuit parameters. This can be seen in its simplest form by comparing Figure 14 (bottom) with Figure 16 (bottom); an additional voltage source in the valve circuit can lengthen the conduction period λ or shorten it.

The application of this method can be observed in Figure 17 (top). Connected in parallel to an inductive valve circuit with an a.c. voltage source v_1 is a second, phase-shifted, a.c. voltage source v_2 through a valve V_2. When thyristor V_2 is triggered the current i_1 is abruptly reduced to zero; V_1 turns off and the current i then flows in valve V_2 and the voltage source v_2.

A further example of current interruption in a valve circuit is illustrated in Figure 17 (bottom). In this case there is no additional voltage source—only an

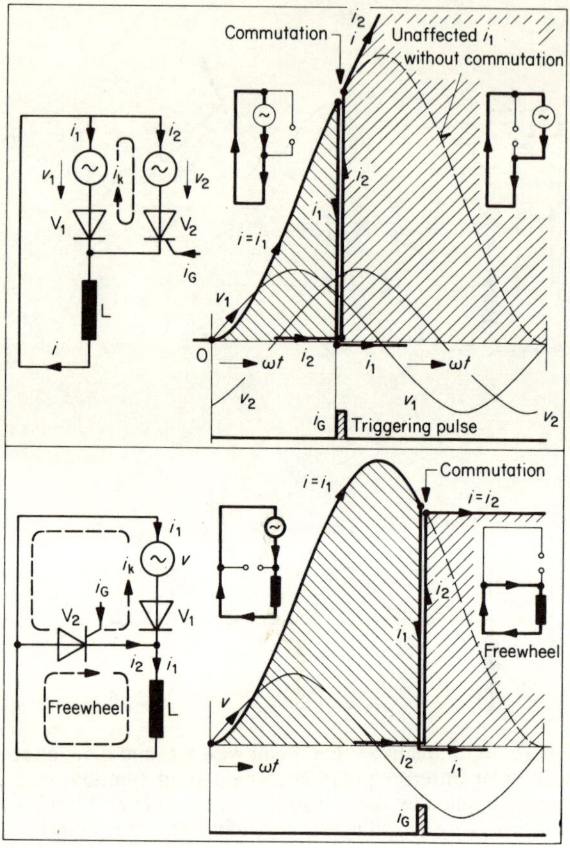

Figure 17. Interruption of current in a valve (V_1) through the triggering of another valve (V_2). The current is transferred—commutated—from valve V_1 to valve V_2. Top: commutation with additional voltage source (v_2); bottom: commutation with subsequent freewheeling. i_k = short-circuit current

additional valve V_2. Following the triggering of valve V_2 and the abrupt termination of current i_1 in valve V_1, valve V_2 continues to conduct and provides a freewheeling path for the inductance L. The current i remains constant at the level i_2 (assuming a lossless circuit).

The valve V_2 can only conduct when a positive voltage is applied to it. This is possible, in the first case (Figure 17, top), only when voltage V_2 is greater than voltage v_1.

In the second case (Figure 17, bottom), V_2 can only be turned on after the supply voltage has reversed.

Thus the current in the conducting valve V_1 can be reduced to zero by turning on V_2, but the latter then takes over conduction of the current. The transfer of current from one valve to another is termed *commutation*. Commutation requires an energy source of sufficient power capacity. In both the cases considered in Figure 17, commutation takes place in an infinitesimal time, and the rise and fall of current in the valves is therefore infinitely fast. For a brief period during commutation, both valves are conducting and the voltage source(s) short-circuited. The valves and the voltage sources form a temporary short circuit, which carries the short-circuit current i_k.

The transfer of current from one valve to another—commutation–forms the basis of operation of all converter circuits.

CAPACITIVE VALVE CIRCUITS

The study of converter circuits is predominantly concerned with valves, inductances, and voltage sources. Valve circuits with capacitance, on the other hand, are important in connection with *protection circuits* and reactive power compensation circuits.

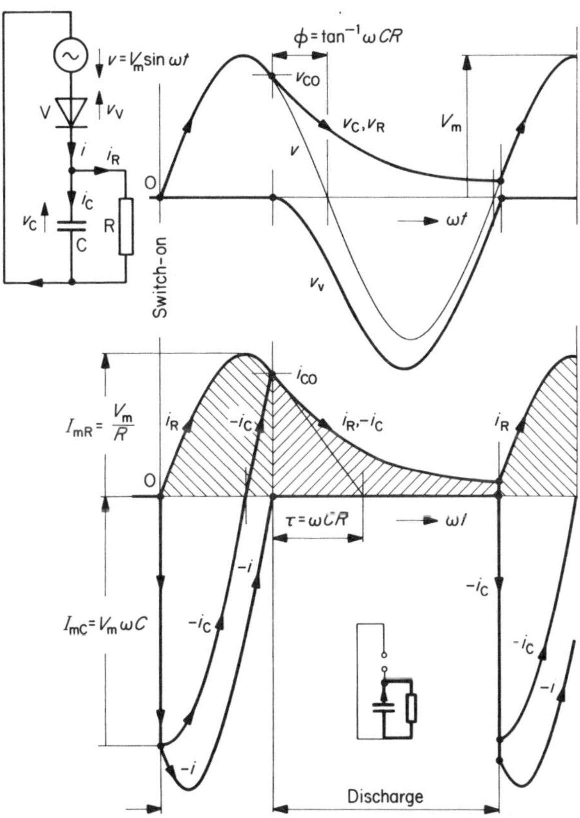

Figure 18. Operation of a valve circuit with capacitor and parallel resistor. Example with a diode and an arbitrary time constant. Currents and voltages in the first cycle after switching on. A continuous current flows in the resistor; the capacitor has a smoothing effect

Figure 18 shows, as an example, voltage waveforms in an a.c. circuit with a diode V, capacitor C, and parallel resistor R. With the voltage v positive, the valve V conducts and the voltage source v is connected directly to the capacitor. The capacitor current leads the voltage by 90°, and the current i_R in the resistor follows the voltage waveform in accordance with Ohm's law. The sum of the two currents represents the current in the valve: $i = i_C + i_R$.

When the current falls to zero the diode blocks. The current zero occurs in advance of the voltage zero by an angle ϕ.

Thereafter the capacitor is decoupled from the voltage source by the blocking diode and discharges through the resistor R with an exponential waveform.

If the circuit is switched on other than at a voltage zero, the capacitor is instantly charged (theoretically by a current pulse of infinite magnitude and infinitesimal duration) to the instantaneous level of the voltage v.

The diode ceases to block when the capacitor voltage reaches the same value as the supply voltage: $v_C = v$.

The valve voltage v_V passes through zero in the positive direction, the valve conducts, and the capacitor is again supplied by the voltage source. The process is repeated, albeit with different initial conditions.

OSCILLATORY CIRCUIT INCLUDING A VALVE

In solving problems associated with overvoltages, filtering, or supply loading, it is often necessary to consider equivalent circuits containing valves, inductances, and capacitors. It is well known that an L–C circuit will begin to oscillate when subjected to a current or voltage excitation: i.e. there arises in it a sinusoidal current whose frequency is determined by the circuit parameters (L, C).

Figure 19 illustrates the behaviour of a simple series tuned circuit following the switching-on, through a thyristor, of a d.c. voltage source E. The most significant result of the process is that the *capacitor voltage* v_C *rises to twice the supply voltage E.*

Since no current can flow in the negative direction, the oscillation is terminated after one half-cycle. The capacitor remains charged to twice the d.c. voltage. The thyristor blocks and is subjected to the difference between the voltages of the d.c. source E and the capacitor v_C; the blocking voltage v_V is equal to the supply voltage E.

MATHEMATICAL APPENDIX III

The natural frequency of an L–C oscillatory circuit is given by

$$f_0 = \frac{1}{2\pi\sqrt{LC}} \tag{14}$$

In the valve circuit of Figure 19 the conduction period of the switching-on pulse is

$$\frac{T}{2} = \pi\sqrt{LC} \tag{15}$$

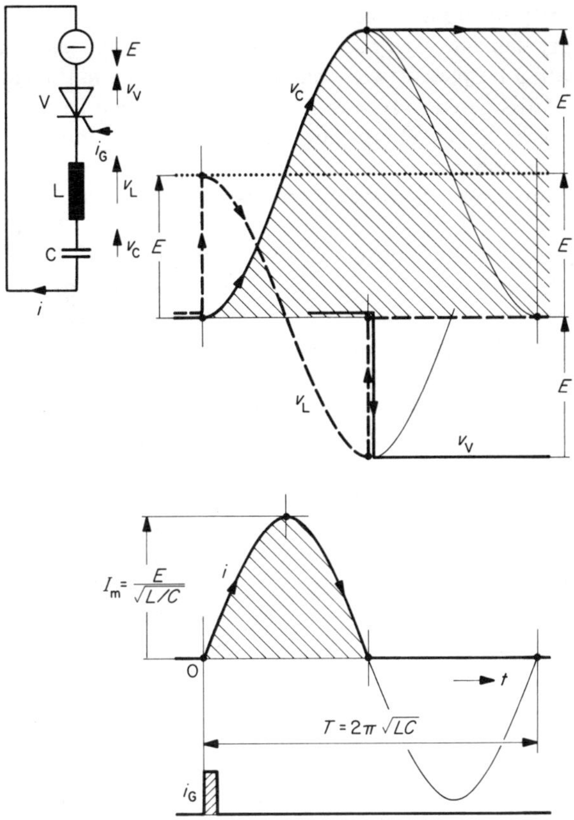

Figure 19. Current and voltage in valve circuit with d.c. voltage source E and a series LC oscillatory circuit. When the thyristor is triggered the capacitance is charged by a single pulse of current i to a voltage equal to twice that of the voltage source

The current amplitude is

$$I_m = \frac{E}{\sqrt{L/C}} \tag{16}$$

The current flows with a sinusoidal waveform

$$i = I_m \sin \omega_0 t \tag{17}$$

POWER BALANCE IN VALVE CIRCUITS

The main purpose of power electronic equipment is the translation of power, in contrast to electronics in general, which is usually concerned with the translation and processing of information.

The power p generally signifies the product of current i and voltage v. *By active power is meant the mean power over a given period of time.*

Figure 20 (left) illustrates as an example the voltages, currents, and powers in an individual element of an a.c. L–R circuit. It will be seen that the power p fluctuates with a frequency twice that of the current i and the voltage v. The active power P dissipated in the resistor R (RI^2) is equal to that supplied by the voltage source ($VI\cos\phi$), and corresponds exactly to the arithmetic mean of the powers p_R and p.

The maximum power available from the supply is termed the apparent power ($S = VI$).

The power that is not utilized is termed the reactive power Q. In the inductance L the power p_L fluctuates symmetrically about zero; the mean value is thus zero and the power may thus be considered as purely reactive power. The amplitude of the power fluctuation is given by the expression $VI\sin\phi$.

In Figure 20 (right) the same circuit is shown (the same voltage v and the same resistance R) as on the left, but the inductance L is replaced by an a.c. regulator—two thyristors connected in inverse parallel, V_A and V_B. The delay angle α is so chosen that the r.m.s. current I, and therefore the active power $P = RI^2$, is the same as in the circuit on the left. The current i is no longer sinusoidal and the power waveforms p, p_R, p_V are quite different from those in the first case. Since the current in and the voltage across the valves do not occur simultaneously, *the valve power p_V is always zero*. The apparent power S from the source and the active power P are the same in both cases. Thus the non-active power Q must also be the same, but it is not 'visible' in the valve

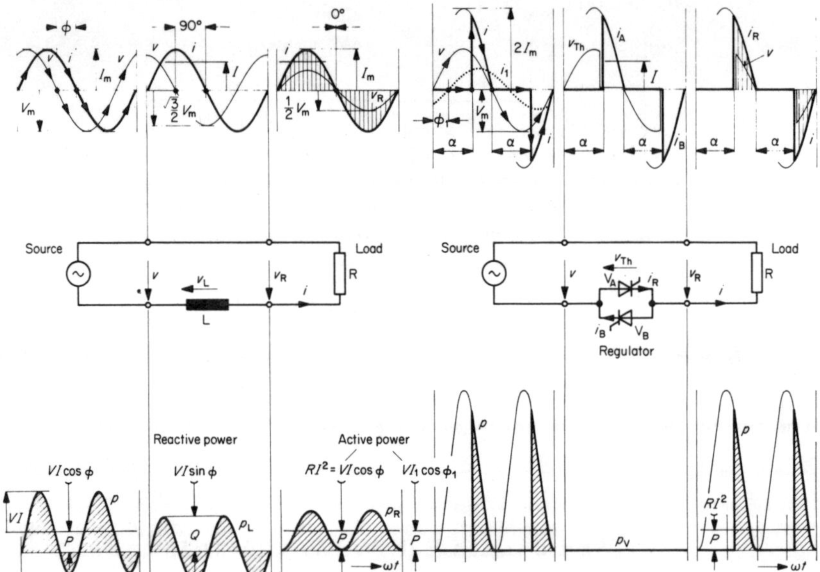

Figure 20. Waveforms of current, voltage, and power in an RL a.c. circuit and a valve circuit (regulator) with two thyristors and ohmic resistance for the same active power P. The power factor is the same in both cases. The power balance is illustrated geometrically in Figure 21. p = instantaneous power

circuit; the valve itself, which is responsible for the non-active power, is free of any power.

In a normal a.c. circuit with sinusoidal waveforms the concept of reactive or non-active power is always associated with the phase shift ϕ. In the case of a non-sinusoidal current, as in Figure 20 (right), it is possible to determine (by Fourier analysis) a sinusoidal component of current (the *fundamental component* i_1) which has the same frequency as the voltage. Thus a phase shift ϕ_1 can be defined between the voltage v and the fundamental component of current i_1, and the normal equations of a.c. circuits applied. Using the trigonometrical functions $\cos \phi_1$ and $\sin \phi_1$, the active power P and the reactive power Q_1 (the *fundamental non-active power*) can be calculated and also the (fundamental) apparent power S_1. This involves only the fundamental component I_1 of the current.

The *total apparent power S,* calculated from the total r.m.s. current I, is, however, greater than the fundamental apparent power S_1, and similarly the total non-active power Q is greater than the fundamental reactive power Q_1, since the total r.m.s. value of the current I is greater than that of the fundamental component I_1. *Any non-sinusoidal current i can be regarded as the sum of sinusoidal currents of different frequencies, amplitudes, and phases. (The mathematical proof is provided by Fourier analysis.)* Thus in addition to the fundamental component i_1, whose frequency corresponds to that of the current i, there are also present the *harmonics* i_n, with n times the frequency. *The r.m.s. current I can be expressed formally as the geometric sum of the fundamental component* I_1 *and the sum of all the higher-frequency components* I_n *(n representing the orders of the harmonics).* A similar process applies, after multiplication by the voltage V, to the powers. As well as the fundamental apparent power S_1, which can be resolved geometrically (by Phythagoras' theorem) into the active power P and the fundamental reactive power Q_1, there is a series of products of the voltage V and the individual harmonics I_n. These powers (VI_2, VI_3,...) do not contribute to the active power flow since the average value of their power–time integrals during a voltage cycle is zero. The volt-amperes associated with the harmonic currents can therefore be considered as *non-active (harmonic or distortion) volt-amperes.* The fundamental reactive power Q_1 and the non-active distortion volt-amperes can then be added geometrically to give the total non-active power.

Although the total non-active power in the two examples of Figure 20 is the same, only part of the non-active power in the valve circuit (right) is due to phase displacement (ϕ_1 is smaller than ϕ). The rest is due to harmonics.

The power balance can be illustrated geometrically. Figure 21 (top) shows the power phasor diagram for the a.c. circuit of Figure 20 (left), while Figure 21 (bottom) shows that for the valve circuit of Figure 20 (right).

The apparent power S could appropriately represent the rated power of the energy source (machine or transformer) that is necessary to provide the required active power. There is, however, no general physical interpretation of non-active power.

In evaluating a circuit in respect of energy, it is useful to introduce the so-called *power factor* λ. representing the active power P as a proportion of

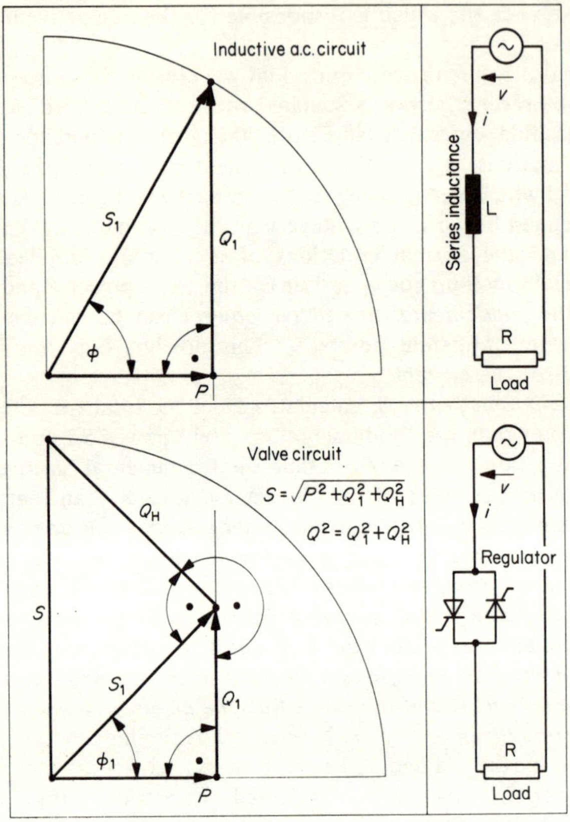

Figure 21. Geometrical representation of the power balance in a valve circuit (bottom) in comparison with that in a simple a.c. circuit (top). The two methods of control, thyristor and series reactor, are equivalent from the point of view of power. The compositions of the non-active powers are, however, different. S is represented by a space diagonal erected on S_1 (see Figure 78)

the apparent power S (P/S). The higher the power factor, the better is the utilization of the energy source.

Figure 22 summarizes the powers in the most important circuit elements—resistance, inductance, d.c. and a.c. voltage sources, and converter valves—with non-sinusoidal current loading. It will be observed that in different components the active power is determined by different current values. In the resistance it is the r.m.s. value (I), in the d.c. source the mean value (I_{AV}), and in the a.c. source the fundamental component (I_1) of the current that gives rise to the active power (P). There is no active power associated with the inductance or the valve.

Although the instantaneous power in the valves is zero at all times, it is possible to determine, by Fourier analysis, fundamental components of current in, and voltage across, the valves which lead to values of active power and fundamental frequency reactive power. The fundamental frequency

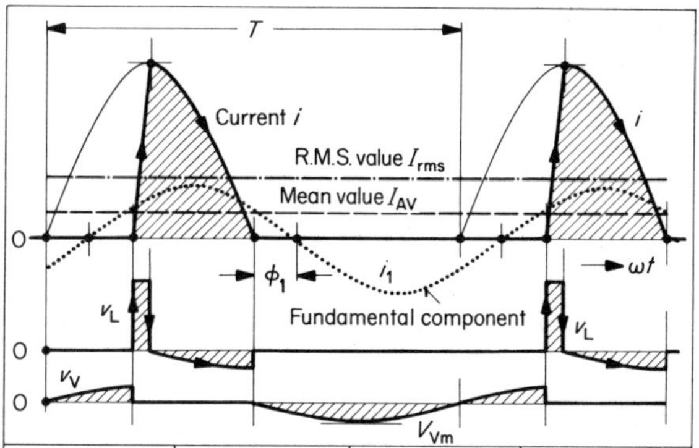

Circuit element	Active power	Non-active power	Apparent power
i, R	$P = RI_{rms}^2$	—	$S = P$
i, L, v_L	—	$Q = S$ $Q_1 = S_1$	$S = V_L I$ $S_1 = \omega L I_1^2$
i, ~, v	$P = VI_1 \cos \phi_1$	$Q = \sqrt{S^2 - P^2}$ $Q_1 = \sqrt{S_1^2 - P^2}$	$S = VI$ $S_1 = VI_1$
i, E	$P = EI_{AV}$	$Q = \sqrt{S^2 - P^2}$ $Q_1 = 0$	$S = EI$
i, v_V	—	—	—

Figure 22. Power equations for the principal circuit elements with non-sinusoidal current (an arbitrary current waveform as an example). The active power is determined by the r.m.s. current in the resistor, by the a.c. voltage source with the fundamental current component, and by the direct voltage with the mean current. $I = I_{rms}$ (root mean square). A fundamental frequency reactive power can also be computed for the valves from the fundamental components of current and voltage

reactive power Q_1 in Figure 21 (bottom) is equal to the product of the fundamental components of the valve current and voltage and their phase displacement ($\sin \phi_1$). In Figure 22, therefore, it would be possible to add a value of reactive power $Q_1 = I_1 V_{V1}$ for the valve.

MATHEMATICAL APPENDIX IV

The instantaneous power in a circuit element is given by

$$p = vi \tag{18}$$

where v = instantaneous voltage across the element
i = instantaneous current in the element

The active power is the arithmetic mean of the power during one cycle:

$$P = \frac{1}{2\pi} \int_0^{2\pi} p \, d\omega t \tag{19}$$

For a sinusoidal waveform,

$$v = v_m \sin \omega t$$
$$i = I_m \sin (\omega t - \phi)$$

The solution of equation (19), after substitution and rearrangement, gives the familiar result

$$P = VI \cos \phi \tag{20}$$

where $v = \frac{1}{\sqrt{2}} V_m$

$$I = \frac{1}{\sqrt{2}} I_m$$

These equations are valid for the calculation of active power only when the voltage and current are sinusoidal and of the same frequency. With non-sinusoidal current the active power from the a.c. source is calculated from the fundamental component of current I_1 and its phase displacement ϕ_1 (Figure 22):

$$P = VI_1 \cos \phi_1$$

The total r.m.s. value of the non-sinusoidal current i is calculated as follows:

$$I = \sqrt{I_{AV}^2 + I_1^2 + I_2^2 + I_3^2 + \cdots} \tag{21}$$

where I_{AV} = arithmetic mean (constant component)
I_1 = fundamental component (r.m.s. value)
$I_2, I_3, \ldots$ = harmonics of the second, third ,..., orders (r.m.s.)

Multiplication by the voltage gives

$$VI = S = \sqrt{V^2 I_{AV}^2 + V^2 I_1^2 + V^2 I_2^2 + V^2 I_3^2 + \cdots} \tag{22}$$

Further,

$$S_1^2 = V^2 I_1^2 = V^2 I_1^2 (\cos^2 \phi_1 + \sin^2 \phi_1) = P^2 + Q_1^2 \tag{23}$$

and

$$Q_H^2 = V^2 (I_{AV}^2 + I_2^2 + I_3^2 + \cdots) \tag{24}$$

so that

$$S = \sqrt{P^2 + Q_1^2 + Q_H^2} \tag{25}$$

where S = total apparent power
P = active power
Q_1 = fundamental reactive power
Q_H = non-active distortion power
S_1 = fundamental apparent power

Geometrically, S in equation (25) corresponds to a three-dimensional diagonal. Thus Q_H is perpendicular to the phasor planes S_1, Q_1, P (shown in one plane in Figure 21).

ENERGY BALANCE

The power–time integral signifies energy. Energy consumption is proportional to active power and to time. No long-term energy flow is involved in non-active power.

In an inductive a.c. circuit (Figure 20, left), the energy in L reverses twice per cycle. It is alternately stored in the magnetic field and released (according to the positive and negative power–time integrals).

Only in the resistance R is energy converted into heat and dissipated. The voltage source v functions also as an energy source.

Energy storage is provided, in general, only by the inductance L and

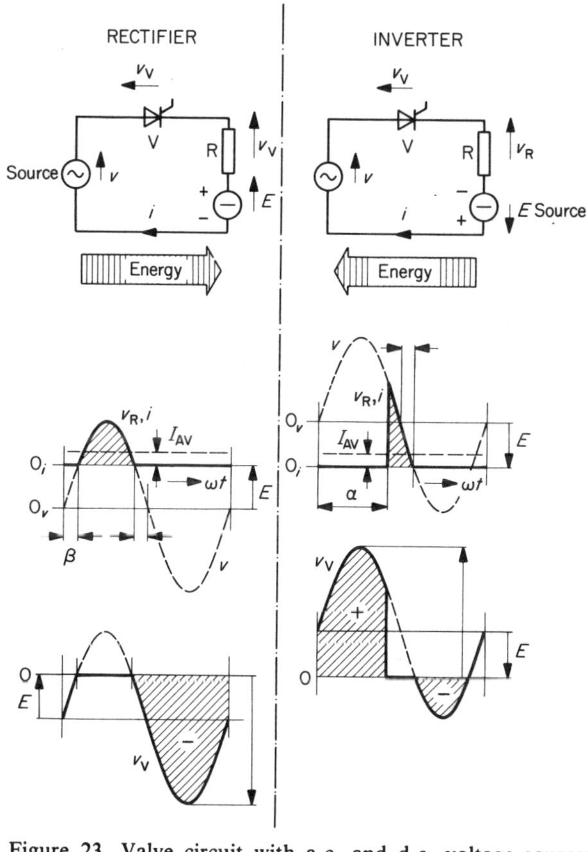

Figure 23. Valve circuit with a.c. and d.c. voltage sources and ohmic resistance. The circuit operates either as a rectifier (left) or as an inverter (right), depending upon the polarity of the d.c. voltage source and the delay angle of the thyristor. The examples show equal active powers

capacitance C. No energy is stored in a valve circuit such as that of Figure 20 (right). The power–time integrals are always positive and there is no reversal of energy flow. The energy consumption in this case is the same as that in the inductive ciruit (left) with pure sinusoidal current. Both methods of control (in Figure 21) are lossless and equivalent from the point of view of energy. Resistance control, on the other hand, results in an increase in the active power drawn from the source, and thus to a greater consumption of energy.

In an ohmic resistance, energy can only be consumed. Inductances, capacitances, and voltage sources can absorb energy and release it. *In a valve circuit with two voltage sources* (v *and* E*, Figure* 15), *it is basically possible for energy to be transmitted in two directions—from the a.c. to the d.c. source or from the d.c. to the a.c. source.* This is illustrated in Figure 23. On the left, the polarity of the d.c. source E is the same as that of the rectified a.c. voltage v, and the difference voltage drives the current i accordingly. The active power $P = EI_{AV}$ has a positive sign (as in resistance R): i.e. the energy is absorbed in the voltage source E. The valve functions as a rectifier and could be uncontrolled (a diode). In Figure 23 (right) the voltage E is reversed. The current depends upon the sum of the two voltages. To obtain the same mean current I_{AV} as on the left, the triggering of the valve V must be delayed ($\alpha > \beta$). The power $P = -EI_{AV}$ is now negative and the d.c. source functions as a source of energy. Although the direction of current has not changed, the transmission of energy is in the reverse direction. The valve in this case has to be controlled (a thyristor) and experiences predominantly positive voltage, in contrast to the previous case, where the valve voltage v_V was negative. The operating mode (in Figure 23, right) is termed 'inversion', since the d.c. energy is converted into a.c. energy.

MATHEMATICAL APPENDIX V

Energy is generally calculated as follows:

$$W = \int p \, dt$$

where p = power

In an a.c. circuit, for a given period,

$$W = PT$$

where P = active power
T = duration

Stored energy is given by:

In an inductance: $W_L = \frac{1}{2} Li^2$ (26)

In a capacitance: $W_C = \frac{1}{2} Cv^2$ (27)

5

The Structure of Multiple-Pulse Converter Circuits

SINGLE-PULSE CONVERTER

A converter valve in an a.c. circuit functions as a *rectifier*. The current can flow only in one direction (Figure 10, bottom) and exhibits a mean value (Figure 13), so that it represents a *direct* current (or unidirectional current). The direct current is in fact discontinuous, with zero periods between pulses. One current pulse occurs in each cycle. The circuit thus constitutes a single-pulse converter. The current flows as a result of one (positive) half-cycle of voltage (the negative half-cycle at the most reduces the current to zero, never increasing it—Figure 11). The utilization of the source is consequently poor.

TWO—PULSE CONVERTER

Figure 24 shows the way in which both half-cycles of the sinusoidal voltage source are utilized. It will be observed that in order to double the pulse number of the single-pulse converter not merely double the number of valves (left) but *four valves* are needed (right). The two-pulse converter thus produced is termed a *bridge*. Figure 25 shows the various ways in which the bridge circuit can be drawn.

The operation of the bridge can be explained with the aid of Figure 26. In the absence of any triggering pulses, all the thyristors of the converter block and no current flows in the load R (period 1). Thyristors 1 and 2 support positive voltage and thyristors 3 and 4 negative voltage. At the electrical angle α (measured from the voltage zero) thyristors 1 and 2 are triggered simultaneously by the pulse of current applied to the gate electrode, and turn on. (The control circuit is omitted for simplicity.) The circuit is completed and a current i_d flows in the load resistance R (period 2). After the current drops to zero, thyristors 1 and 2 block, and no current flows in the following period 3. Thyristors 3 and 4 now support positive voltage and can therefore be triggered at angle α (again measured from the voltage zero). In the following period (4), thyristors 3 and 4 carry current, thyristors 1 and 2 support negative voltage, and the same current flows in the load as in the first half-cycle. In the second cycle the whole process is repeated.

SIX-PULSE (THREE-PHASE) CONVERTER

The transmission and distribution of electrical energy is based upon the three-phase system, and most industrial equipment is supplied from three-phase

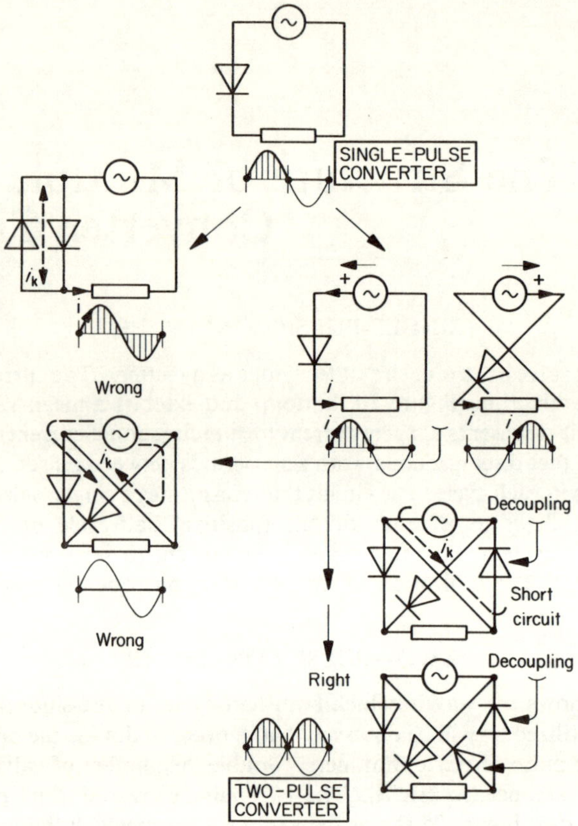

Figure 24. Development of the two-pulse converter (single-phase bridge). To double the pulse number, in comparison with the single-pulse converter, four times as many valves are required (right). With two valves (left) the result is either an a.c. regulator or a short circuit

mains. It follows that the three-phase system is also preferred for converters (particularly in the high-power range).

The three-phase system is formed by three voltages displaced in phase by 120 ° el. A logical extension of the single-phase bridge circuit—from two a.c. input lines to three—leads to the *three-phase bridge*. This is illustrated in Figure 27. (It is also possible to use three single-phase bridges; this, however, requires twice as many valves.) The three-phase bridge (bottom) consists of six converter valves, the single-phase bridge (middle) has four valves, and the single-pulse converter (top) one. The maximum obtainable value of the converter voltage—the *ideal d.c. voltage* V_{dio}—is, in the case of the single-phase bridge, twice, and in the case of the three-phase bridge three times, that of the single-pulse converter. In terms of pulse number, however, the three-phase bridge shows a disproportionate increase over the single-phase bridge—the d.c. voltage exhibits six pulses in each cycle.

The three-phase bridge is usually connected to the distribution network

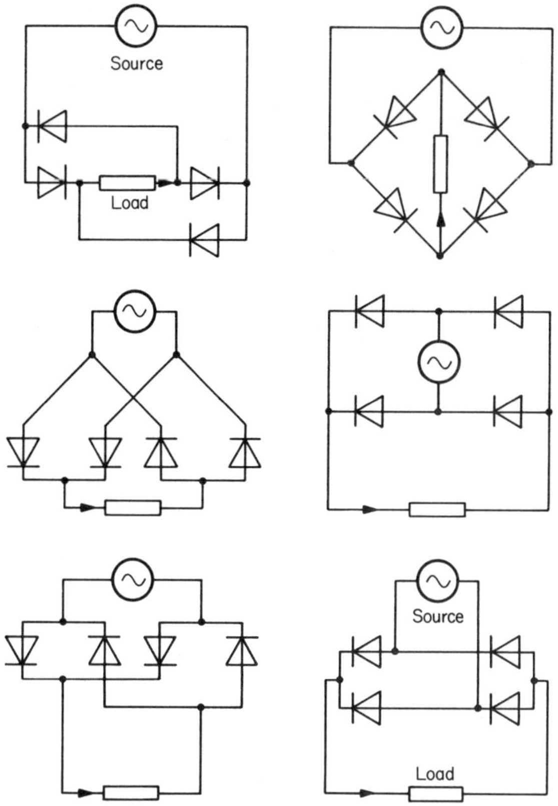

Figure 25. Principal circuit diagrams of the single-phase bridge converter. The converter circuit can be drawn in many different ways. The diagrammatic representation to be chosen is the one most suitable for the purpose

through a matching transformer. The *converter transformer* therefore represents an important component of the converter installation, and the construction of the transformer and the specification of the current and voltage ratings consequently form a part of converter circuit theory.

In a three-phase bridge circuit the converter is connected to the secondary winding of the transformer. The transformer secondary can be connected either in star or in delta (Figure 28). The operation of the converter is not affected by the transformer connection—the d.c. voltage (V_d) is always related to the line-to-line voltage.

The valves of the three-phase bridge may be either diodes or thyristors.

MATHEMATICAL APPENDIX VI

Assuming an ideal supply, the mean value of the d.c. converter voltage, on no-load and in the condition of maximum output (diode operation), is represented as the *ideal d.c. voltage* V_{dio}.

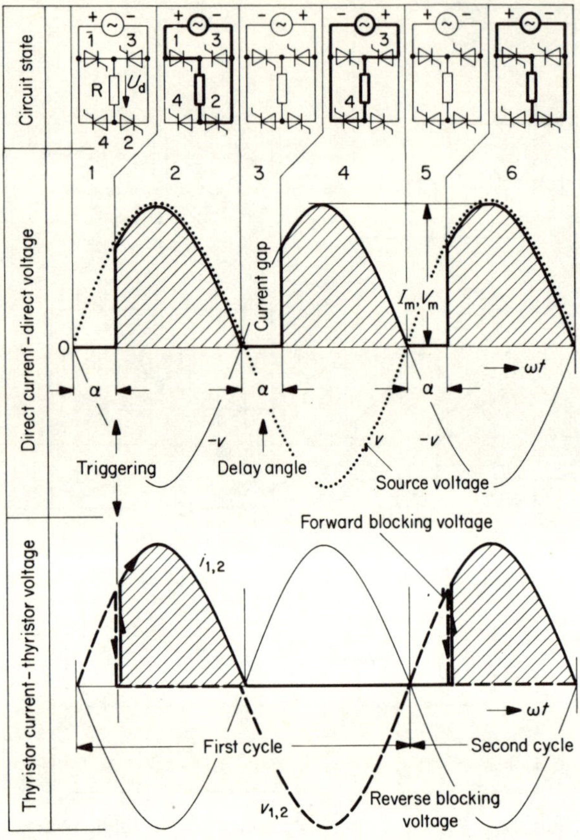

Figure 26. Operation of an ideal fully controlled single-phase bridge with resistive load. In one cycle the circuit state changes four times. Both half-cycles of the supply voltage are utilized. The direct current $i_d = v_d/R$ is discontinuous

In the *single-pulse converter circuit,*

$$V_{dio} = \frac{1}{2\pi} V_m \int_0^{180°} \sin \omega t \, d\omega t = \frac{1}{\pi} V_m = 0.318 V_m \tag{28}$$

In the *two-pulse converter,*

$$V_{dio} = \frac{1}{\pi} V_m \int_0^{180°} \sin \omega t \, d\omega t = \frac{2}{\pi} V_m = 0.637 V_m \tag{29}$$

The *six-pulse three-phase bridge* the ideal d.c. voltage is given by

$$V_{dio} = \frac{1}{\pi/3} V_m \int_{60°}^{120°} \sin \omega t \, d\omega t = \frac{3}{\pi} V_m = 0.955 V_m \tag{30}$$

The corresponding waveforms of the d.c. voltage and the values of V_{dio} are shown in Figure 27.

The ideal d.c. voltage V_{dio} is one of the most significant parameters of the converter circuit.

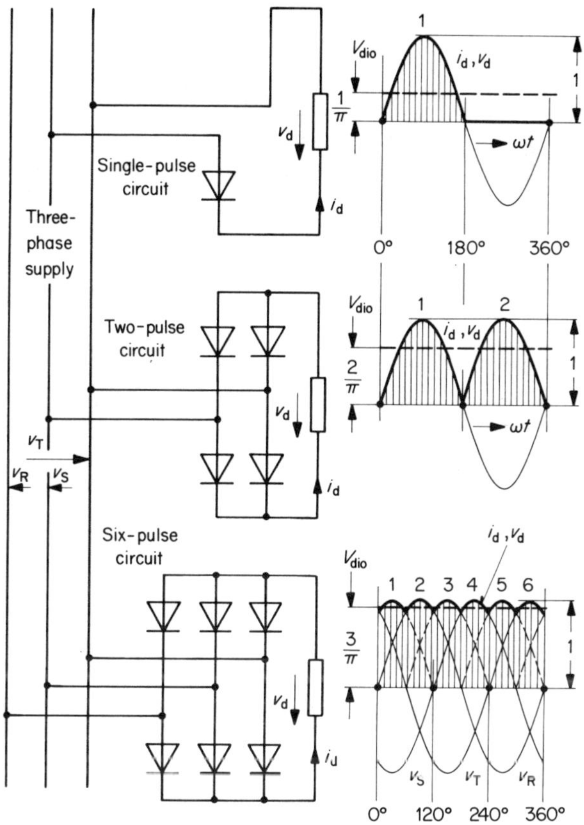

Figure 27. Basic converter circuits connected to a three-phase supply. Logical extension of the single-phase (two-pulse) bridge (middle) leads to the three-phase (six-pulse) bridge (bottom). The direct voltage (direct current)—right—applies to diode operation, resistive load, and ideal voltage sources. V_{dio} = ideal no-load direct voltage

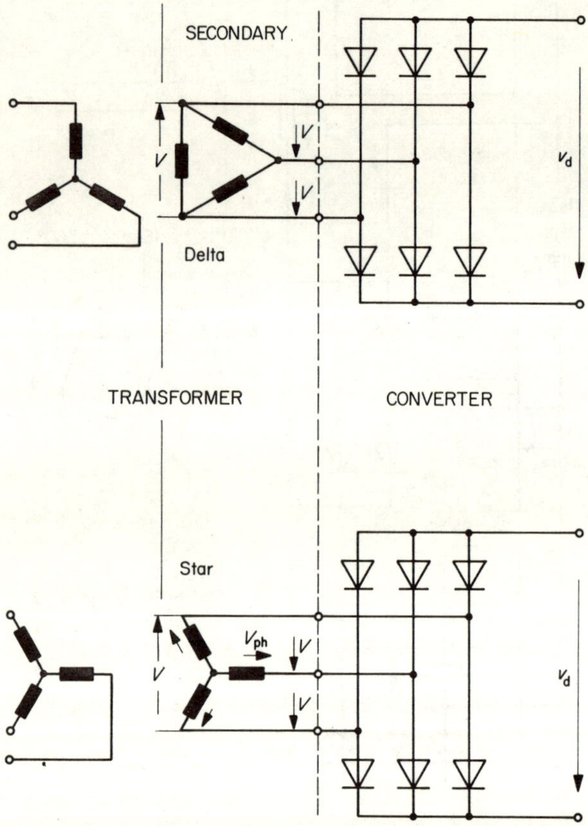

Figure 28. The three-phase bridge converter is usually fed from a matching transformer. The secondary windings of the transformer can be connected in delta (top) or in star (bottom). In either case the direct voltage V_d is derived from the line voltage V

6

Operation of the Three-Phase Bridge Circuit

UNCONTROLLED THREE-PHASE BRIDGE WITH IDEAL SUPPLY

The basic mode of operation of the three-phase bridge circuit can be illustrated with reference to Figure 29. The converter is uncontrolled, using diodes. It is assumed that the load is an ohmic resistance and the supply an ideal voltage source. In the first period the voltage v_S is positive with respect to the forward direction of diodes 1 and 5, so that these diodes are conducting. All the other diodes are blocking. The voltage v_S is thus applied to the load resistance. In the second period the voltage v_R attains the highest value, and since it is positive with respect to diodes 1 and 6, the current is now carried by these. The previously conducting diode 5 blocks. The load resistance is supplied from the

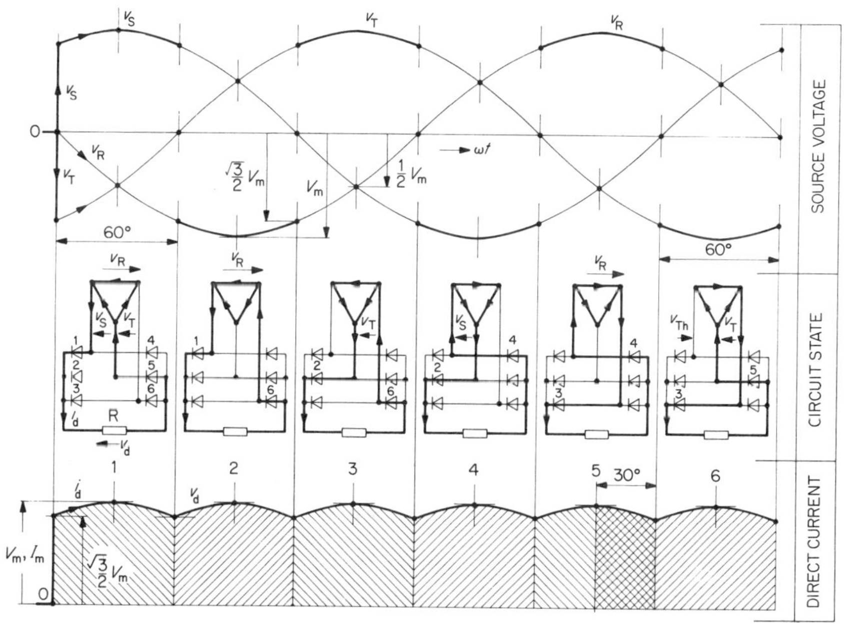

Figure 29. Operation of the three-phase bridge. Circuit states and waveform of direct current i_d (and direct voltage v_d) for resistive load, ideal voltage sources, and diodes. The converter operates in a six-pulse mode: the direct voltage exhibits six equal pulses in one cycle

voltage v_R. The direction of current in the load does not change and the current i_d repeats the waveform of the first period. Each period lasts 60° el—i.e. one-sixth of a cycle—so that *in one cycle the circuit passes through six different switching conditions.* There are always two diodes carrying current simultaneously. Each diode conducts for two consecutive periods. The conduction period of the valve is therefore 120° el (or one-third of a cycle). In the load, the current waveform is the same in every period. The d.c. voltage v_d is composed of an envelope of the voltage peaks. Since the load is resistive, the voltage waveform v_d is similar to that of the current i_d.

In one cycle the d.c. voltage exhibits six pulses. *The circuit operates in a six-pulse mode.* The converter functions as a six-pulse rectifier.

Figure 30 shows the currents and voltages in an individual diode. Each valve conducts for one-third and blocks for two-thirds of a cycle.

Figure 29 shows how the switching conditions in the converter change during a complete operating cycle (one cycle of the supply). The conducting

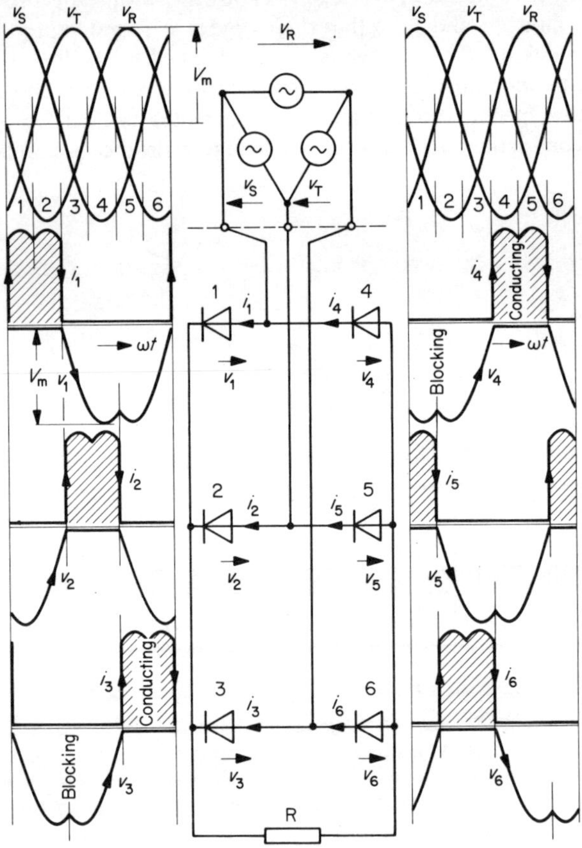

Figure 30. Valve currents and valve voltages in the three-phase bridge. The converter is furnished with diodes. Resistive load and ideal three-phase system. Each diode conducts for one-third of the cycle and blocks for two-thirds

path at any particular time is drawn in heavy lines. Every circuit condition is, however, a variation on the same basic pattern: two valves are conducting, two valves support the full voltage of the supply, and two valves are in series relative to the d.c. side. This basic pattern is illustrated in Figure 31. The valve and source positions are indicated by letters. In each period these positions are held by a new configuration of valves and source (as indicated in the table). Thus, according to this illustration, the circuit remains unaltered during the operating cycle: only the individual elements change their positions.

The rectifying action of the three-phase bridge can also be effected by a mechanical commutator (the contact rectifier). The switching logic is illustrated in Figure 32. The switching has to take place in the same sequence as has been described above. With diodes, *the supply mains itself ensures the correct switching sequence. The converter is 'supply-commutating'.* Rectification occurs without any external influence or control.

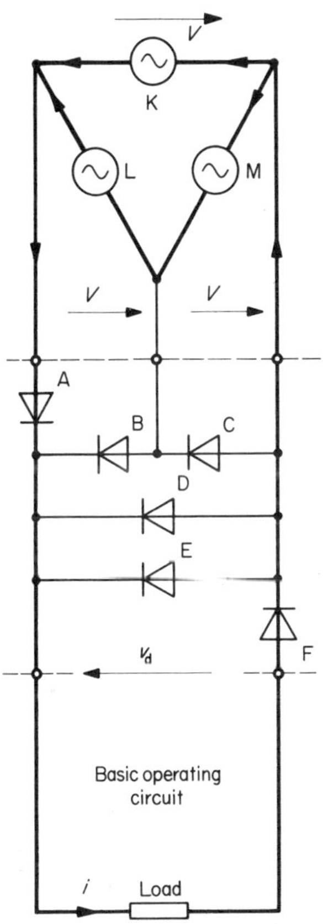

Position	Voltage source assignment					
K	+S	−R	+T	−S	+R	−T
L	+R	S	S	T	T	R
M	+T	T	R	R	S	S
Period	1	2	3	4	5	6
A	1	1	2	2	3	3
B	3	2	1	3	2	1
C	6	5	4	6	5	4
D	4	4	5	5	6	6
E	2	3	3	1	1	2
F	5	6	6	4	4	5
	Valve assignment					

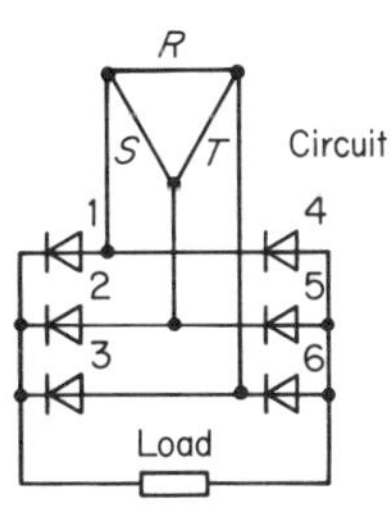

Figure 31. Basic circuit state of the ideal three-phase bridge. During one operating (supply) cycle the valves and the voltage sources change their positions in the circuit six times in accordance with the assignment table. (Compare the illustration in Figure 29)

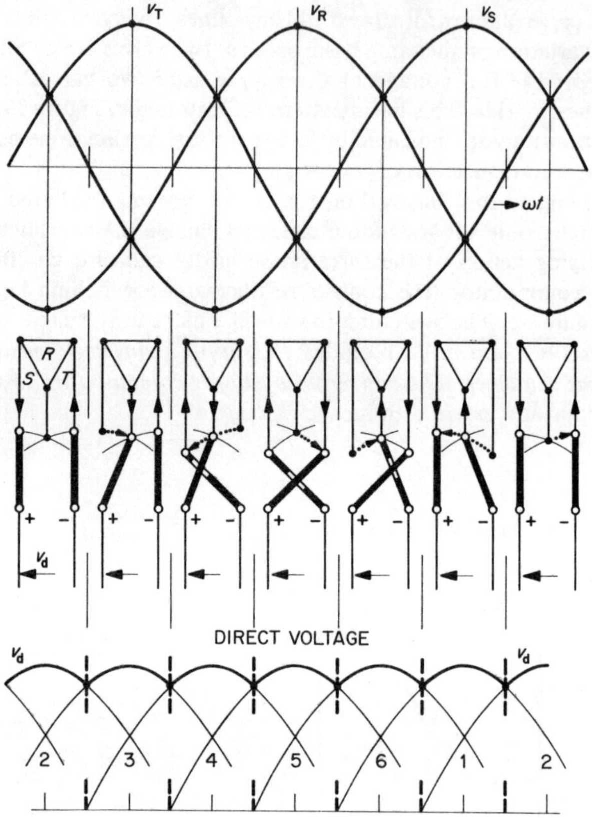

Figure 32. Operation of the three-phase bridge. Switching logic in the circuit state variations. Rectifier operation can be achieved with mechanical switches as shown. With diodes, switching in the correct sequence is brought about by the supply voltage

MATHEMATICAL APPENDIX VII

The ideal d.c. voltage of the three-phase bridge, from equation (30), is

$$V_{\text{dio}} = \frac{3}{\pi} V_{\text{m}} \tag{31}$$

With resistive load the mean direct current is

$$I_{\text{do}} = \frac{3}{\pi} I_{\text{m}} \qquad \text{where } I_{\text{m}} = \frac{V_{\text{m}}}{R} \tag{32}$$

The r.m.s. value of the direct current is calculated as follows (see Figure 29, period 5):

$$I_{\text{dorms}} = I_{\text{m}} \sqrt{\frac{1}{\pi/6} \int_0^{30^\circ} \cos^2 \omega t \, \mathrm{d}\omega t}$$

$$= I_{\text{m}} \sqrt{\frac{3}{2\pi}} \sqrt{\frac{\pi}{3} + \frac{\sqrt{3}}{2}} = 0.9558 I_{\text{m}} \tag{33}$$

The mean valve current is obtained from the direct current:

$$I_{\mathrm{VAV}} = \frac{1}{3} I_{\mathrm{do}} \tag{34}$$

since the valve conducts for one-third of the cycle. The r.m.s. value of the valve current is then

$$I_{\mathrm{Vrms}} = \frac{1}{\sqrt{3}} I_{\mathrm{do}} \tag{35}$$

The maximum valve blocking voltage is the peak value of the source voltage

$$V_{\mathrm{Vm}} = V_{\mathrm{m}} \tag{36}$$

The r.m.s. valve blocking voltage is

$$V_{\mathrm{Vrms}} = 0.633\, V_{\mathrm{m}}$$

UNCONTROLLED THREE-PHASE BRIDGE WITH INDUCTIVE SUPPLY

Equivalent circuit

For certain basic investigations of converter operation it is useful to regard the three-phase supply system as ideal (see Figures 27, 29, 30, and 31). To obtain a realistic insight into the characteristics of the converter and to understand the behaviour of the circuit under different operating conditions it is, however, essential to take into account the inductance of the system (resistances and capacitances can be neglected).

Usually the three-phase supply for the converter is provided through a *transformer*. In that case the supply mains can be regarded as ideal (with infinite short-circuit capacity). It is the short-circuit characteristics of the transformer that are significant in the study and understanding of the converter. If a transformer short-circuited on the secondary side is connected to the supply mains, a current I_{k} flows in the short circuit, in the steady state, which is much greater than the rated current. The short-circuited transformer presents itself to the supply as an inductance L_{k} (with reactance ωL_{k}). This short-circuit inductance is calculated from the ratio of the secondary voltage V_{sec} to the short-circuit current I_{k} (which, according to the impedance law, gives the reactance $X_{\mathrm{k}} = \omega L_{\mathrm{k}}$).

Every transformer has also its specified short-circuit voltage e_{k}. The short-circuit voltage e_{k} is a primary voltage expressed in terms of the rated primary voltage V_{p}. If the transformer with secondary short circuit is supplied with this voltage on the primary side, the current loading of the transformer corresponds exactly to the rated current I_{N}.

The short-circuit current I_{k} of the transformer, relative to the rated current I_{N}, is equal to the reciprocal of the corresponding short-circuit voltage e_{k}. (For example, if $e_{\mathrm{k}} = 5$ per cent., $I_{\mathrm{k}} = 20 I_{\mathrm{N}}$.) The transformer's short-circuit reactance X_{k} is directly proportional to the short-circuit voltage e_{k}; the larger the short-circuit voltage e_k, the smaller is the short-circuit current I_k and the greater is the reactance X_k.

The *short-circuit inductances* L_k of the transformer are taken into account in the derivation of the equivalent circuit of the three-phase system for the purposes of converter circuit theory. This is shown in Figure 33. The three-phase supply system here consists of three ideal voltage sources, corresponding in value to the rated secondary voltage of the transformer, and three equal reactances $X_k = \omega L_k$. Only the short-circuit reactance of the transformer is considered; the no-load reactance, like the ohmic resistance, is neglected. It is further assumed that the reactances X_k are linear, and that they are the same for a three-pole as for a two-pole short circuit. (For the effect of supply reactances, see Figures 116 and 117.) In the d.c. circuit it will be appropriate hereafter to assume a load consisting of a d.c. source E and an inductance L.

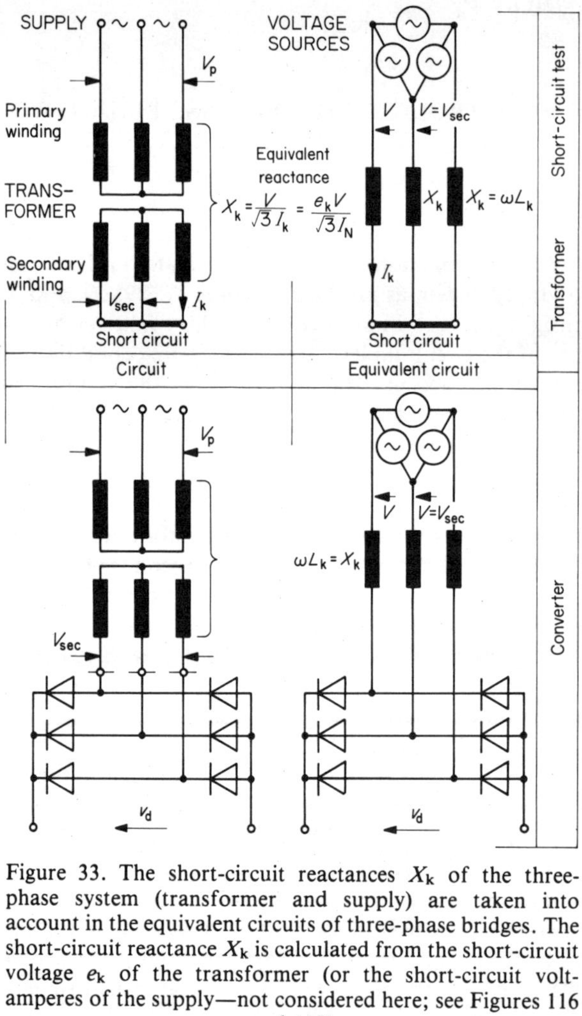

Figure 33. The short-circuit reactances X_k of the three-phase system (transformer and supply) are taken into account in the equivalent circuits of three-phase bridges. The short-circuit reactance X_k is calculated from the short-circuit voltage e_k of the transformer (or the short-circuit volt-amperes of the supply—not considered here; see Figures 116 and 117)

Discontinuous current operation with high load back-voltage

If the load back-voltage E is greater than the amplitude of the sinusoidal a.c. voltage V_m, no current can flow in the converter and the diodes block continuously. This can be seen in Figure 34 (left).

If the voltage E is less than the amplitude V_m, two different operating conditions can arise. If the conduction angle λ is less than 60°, the three-phase bridge behaves like six independent single-pulse converters. The current waveform can be constructed simply, in accordance with the graphical method illustrated in Figure 15. The direct current i_d is composed of individual pulses separated by gaps. If, on the other hand, the current conduction angle exceeds 60°, the diode currents overlap. The direct current is continuous (Figure 34, right) and assumes a steady value after a certain delay following the voltage change.

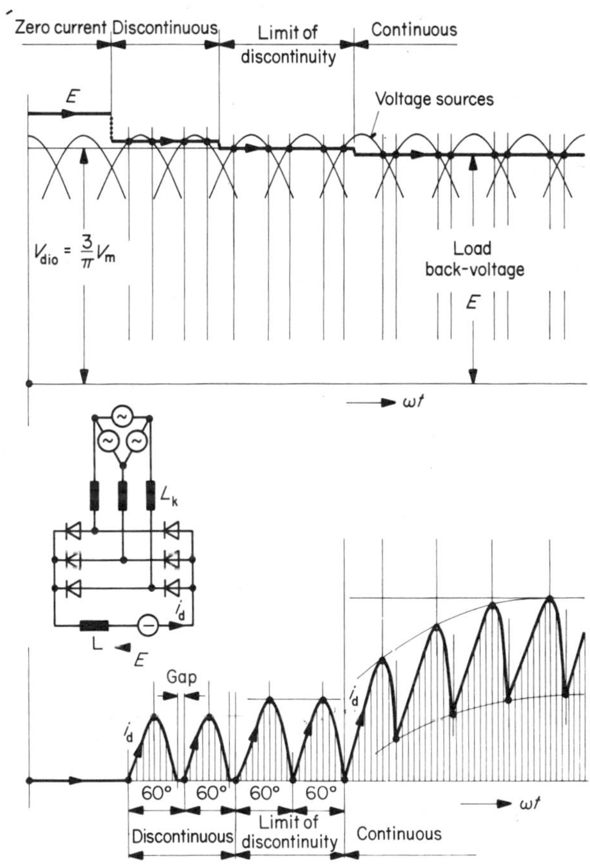

Figure 34. Uncontrolled three-phase bridge connected to a supply with inductance (L_k). Load: back-voltage E and inductance L ($L < L_k$). Waveform of direct current i_d with three different back-voltages E. In the range of voltages $1 < E/V_{dio} < \pi/3$, i_d is discontinuous

Formation of continuous direct current—commutation

If the load back-voltage E is less than the open-circuit d.c. voltage $V_{dio} = (3/\pi)V_m$, the current does not fall to zero after 60° and the next diode (in the switching sequence of the converter) switches onto valves that are still conducting. Figure 35 (left) shows the current waveform after the converter is switched on (the switching corresponding to period 2 of Figure 29). The relative circuit and the equivalent circuit are shown in Figure 36 (left). The current waveform can be constructed, in accordance with the principle of superposition, as the difference between the a.c. and the d.c. component, $i_d = i_\sim - I_E$ (Figure 15).

The effective circuit inductance after switching on consists of the sum of two short-circuit inductances $2L_k$ and the load inductance. Diodes 1 and 6 conduct initially. The current ($i_d = i_1 = i_6$), driven by the voltage v_R, increases. Across each inductance L_k appears a voltage drop v_{Lk}. The d.c. voltage is therefore less than the source voltage by $2v_{Lk}$. The valve blocking voltage v_2 is less than the source voltage by v_{Lk}, and therefore reaches zero earlier, by an angle β, than it would in the absence of L_k. At the instant when $i_1 = i_0$ diode 2 turns on.

The equivalent circuit of Figure 36 (right) now applies. Two phases of the transformer are short-circuited through the simultaneously conducting diodes

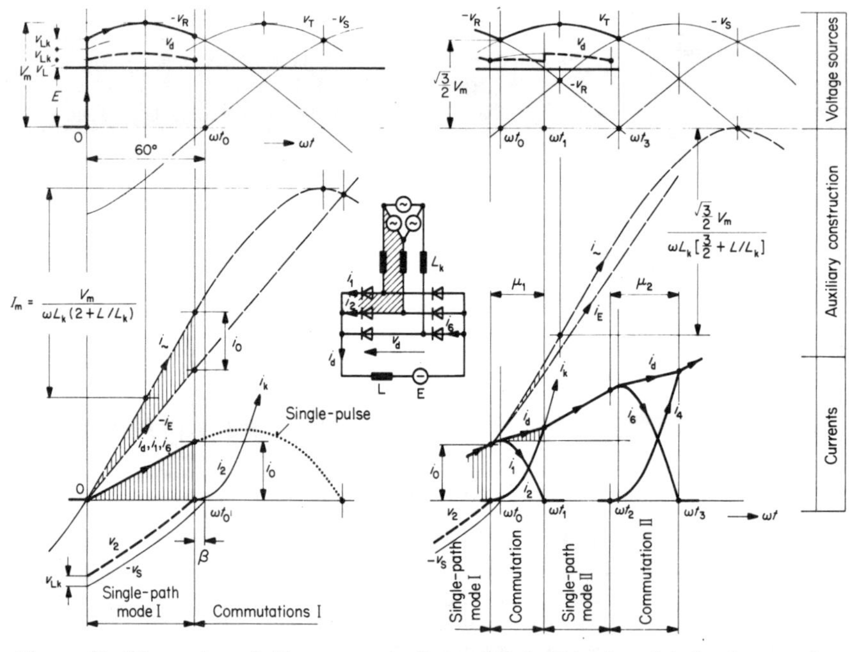

Figure 35. Three-phase bridge, uncontrolled, with L–E load and inductive supply. Switching-on process with continuous direct current. Graphical construction of the current waveform for the single-path mode (left) and for commutation (right), according to the equivalent circuit of Figure 36. It is apparent that an analytical solution for the converter currents is quite complicated. Most studies, including those on which many of the diagrams in this book are based, are carried out with the aid of computers

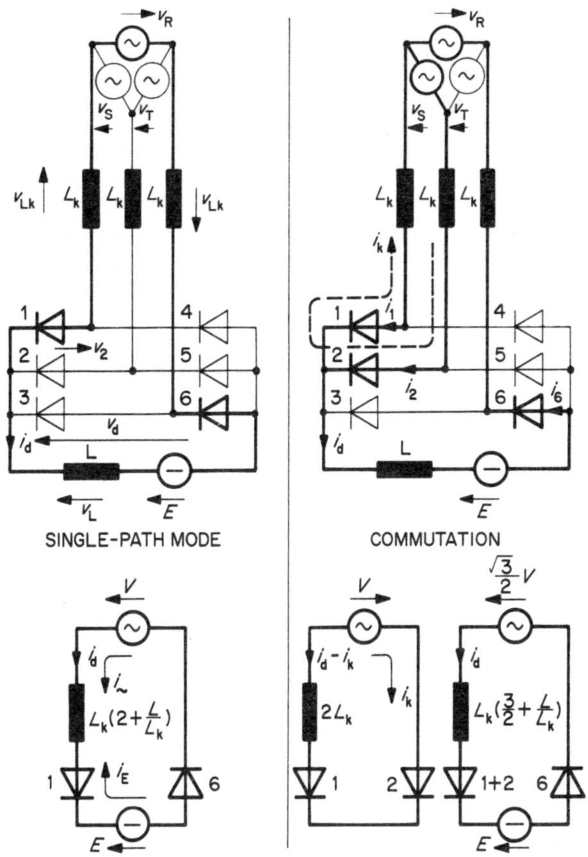

Figure 36. Equivalent circuits for calculation in the three-phase bridge circuit for the single-path mode and for commutation. The corresponding current and voltage waveforms are reproduced in Figure 35. During one cycle the circuit state of the converter changes twelve times

1 and 2. The (short-circuit) current i_k increases in valve 2 and, since it opposes the valve current i_1, the latter decreases. The current i_1, consisting of the difference between i_d and i_k ($i_1 = i_d - i_k$), falls and reaches zero at angle ωt_1. The current i_k is driven by the voltage v_s and limited by two short-circuit inductances $2L_k$. The load current i_d, on the other hand, is driven only by a voltage $0.5\sqrt{3}V - E$ (because of the two-phase short circuit) and limited by the effective inductance $(L + 1.5L_k)$(two inductances L_k are connected in parallel through diodes 1 and 2, resulting in a value of only $L_k/2$).

At ωt_1 diode 1 blocks and conduction is assumed by diode 2. Diode 6 continues to conduct. The converter remains in this condition (3 in Figure 29) until ωt_2 is reached; diode 4 then turns on. The transformer is again short-circuited across two phases through diode 4 and 6, the current in valve 6 falls, and at ωt_3 diode 4 assumes the total flow of current.

The transfer of current from one valve of the converter to another is termed

'commutation'. During commutation three valves are conducting. The period of commutation is defined as a commutation angle μ (the overlap angle, in electrical degrees).

In an ideal circuit without inductance, commutation takes place in an infinitesimal time ($\mu = 0$). The circuit conditions then change instantaneously (Figure 29).

Figure 35 (right) illustrates the first two commutations following the connection of the three-phase bridge to the supply (left). The current i_d increases progressively and reaches its steady-state value only after a certain time interval, which depends mainly upon the ratio L_k/L (Figure 37).

7

Theory of the Three-Phase Bridge

ASSUMPTIONS

From the example of the operation of the three-phase bridge circuit with L–E load and commutating inductance L_k (Figures 35 and 36) it can be seen that an exact calculation of the currents and voltages becomes quite complicated. For this reason it is usually assumed in converter circuit theory that *the load inductance L is very high* ($L >> L_k$) or, what amounts to the same thing, that the direct current I_d is completely smooth, an assumption which simplifies the mathematics of converter calculations considerably. This generally adopted theoretical simplification might be termed 'conventional theory'.

Figure 37 shows the waveform of direct current i_d in the three-phase bridge after switching on for three different inductance ratios. The greater the ratio of the inductances L/L_k, the longer is the duration of the transient switch-on process. At the same time, the ripple in the direct current, $\Delta i_d/I_d$, is reduced. However, the mean value of the direct current in the steady state is not affected by the load inductance. With a large inductance ratio L/L_k the direct current i_d can be assumed to be smooth. For this case, as will be shown, it is possible to establish simple formulae and equations for the three-phase bridge.

The conventional theory cannot, however, be applied in the limiting cases (discontinuous current, variable delay angle, etc.), but must be combined appropriately with other assumptions (see Figure 70).

UNCONTROLLED THREE-PHASE BRIDGE

Simple commutation—typical operating mode of the three-phase bridge

Figure 38 shows the valve currents and the direct voltage v_d in the uncontrolled three-phase bridge in the steady-state condition with smooth direct current. At the bottom of the diagram the corresponding switching states of the converter are illustrated. It can be seen that *the switching state of the converter alternates*. Either two valves ('single-path' mode) or three ('commutation') conduct simultaneously. During commutation one valve (e.g. 6 or 2) takes over the current from another (5 or 1). The direct current i_d remains constant. There is thus no voltage developed across the commutating inductance L_k in the single-path mode ($di/dt = 0$). Only during the commutation interval μ is there a voltage drop v_{Lk} across the inductance L_k due to the current variation. The commutating current i_k is related to the short-circuit current of the transformer. During commutation there is in fact a limited short circuit across

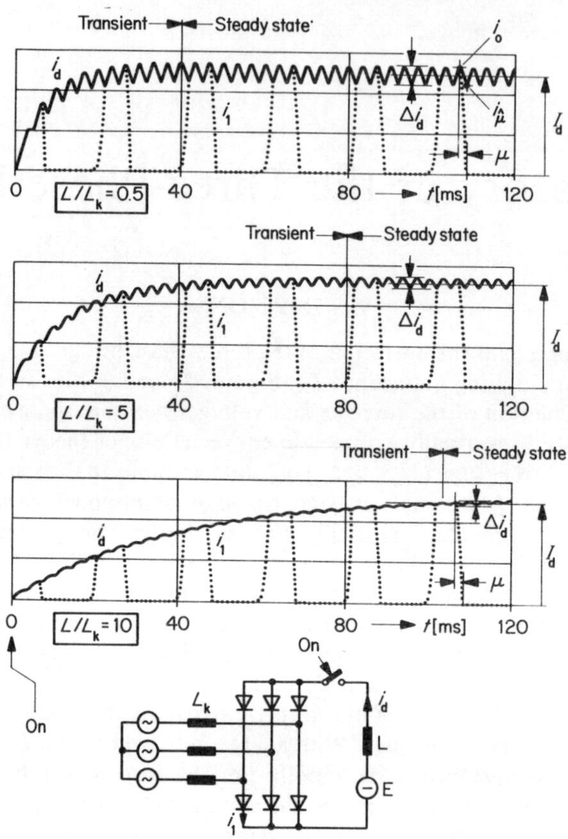

Figure 37. Switching-on process of the three-phase bridge with L–E load and commutating inductances L_k. Effect of the load inductance on the build-up time (transient state) and the ripple current (Δi_d). With a large inductance ratio $L/L_k > 10$ the direct current i_d can be considered smooth

the transformer (or supply mains), although the short-circuit current i_k rises only to the level of the direct current I_d.

In the steady-state condition the direct voltage V_d (mean value) is equal to the back-voltage E. During the commutation interval (the overlap angle μ) the direct voltage v_d follows a sine wave with amplitude $(\sqrt{3}/2)V_m$. In the single-path mode the instantaneous direct voltage v_d is equal to the highest of the source voltages. The mean direct voltage V_d is reduced from the ideal no-load value $V_{dio} = (3/\pi)V_m$ by a voltage drop V_{dx}. *The converter thus behaves like a voltage source V_{dio} with an internal resistance:* the direct voltage V_d falls as the current I_d increases.

In the course of one period the switching state of the converter changes altogether twelve times. Each commutation is followed by a state without commutation (single-path mode).

Figure 38 shows only a short period of time. The way in which the switching state of the three-phase bridge changes during the whole supply cycle is illus-

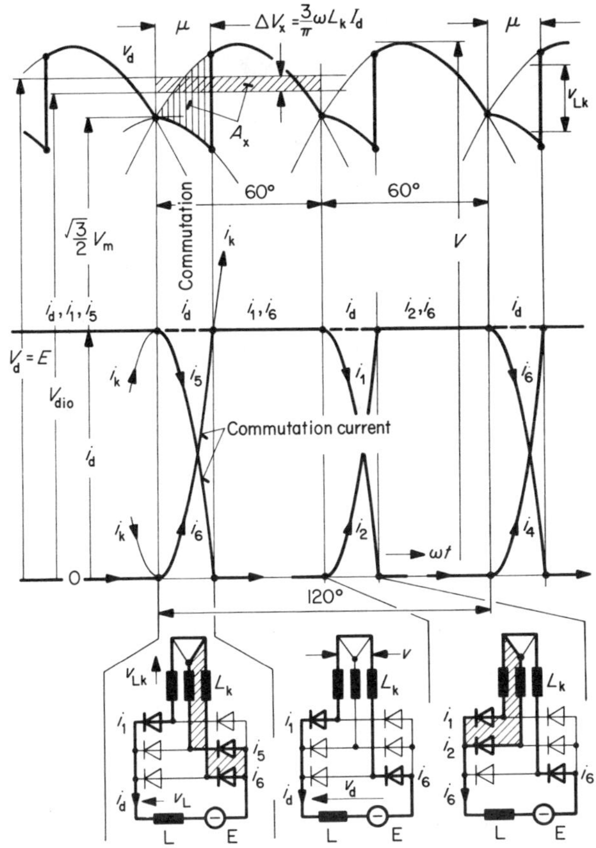

Figure 38. Conventional representation of the currents and the direct voltage in the three-phase diode bridge. Period: half-cycle. Direct current assumed smooth ($L \gg L_K$). During a commutation the transfer of current takes place from one valve (5) to another (6). The transformer is thereby short-circuited between two phases

trated graphically in Figure 39. This shows the complete switching cycle. The conducting states of the valves are indicated in the middle of the diagram. From Figure 39 the following principles can be deduced:

1. *Six commutations occur in one cycle.* Each commutation is followed by a state without commutation, so that in total the converter exhibits twelve distinct switching states.
2. The diodes on the positive side of the d.c. circuit (1, 2, 3) commutate alternately with those on the negative side (4, 5, 6). It is possible therefore to speak of two independent 'commutating groups' operating 180° out of phase.

Every switching state of the three-phase bridge of Figure 39 can be represented by one of the two basic circuits shown in Figure 40. In this illus-

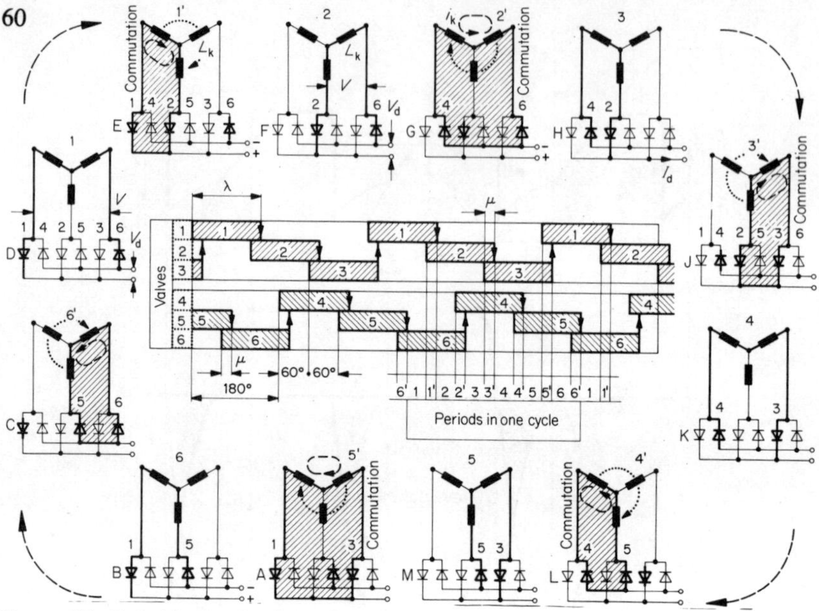

Figure 39. Switching cycle of the three-phase bridge in the operating range of (simple) commutation. Circuit states and pattern of valve conduction periods (λ). Transformer secondary windings star-connected. Load not shown. During commutation three valves conduct simultaneously

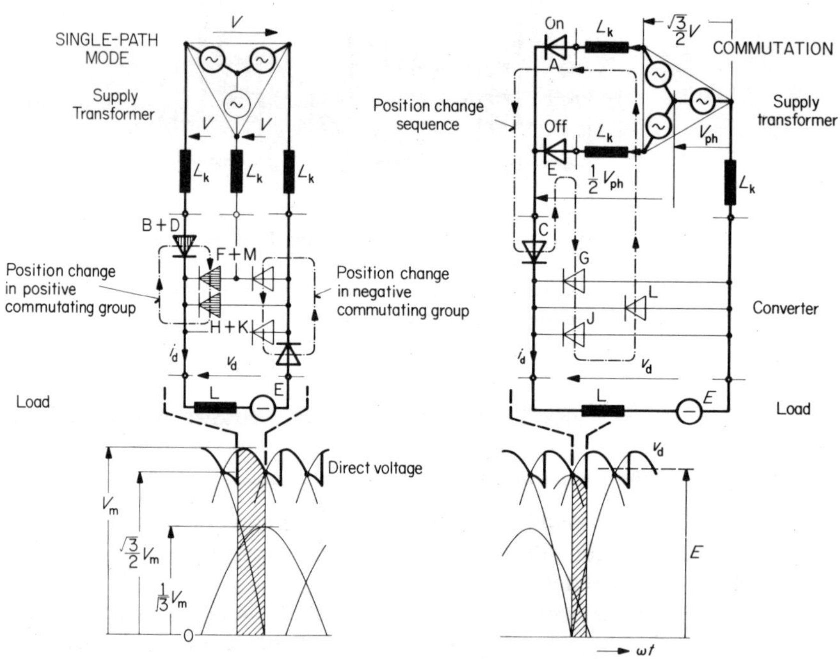

Figure 40. Basic circuit state of the three-phase bridge in periods without commutation (single-path mode: cf. Figure 31) and with commutation (right). Every circuit state of the converter (Figure 39) can be represented by one of these two basic circuits. The position–change sequence corresponds to the letter sequence from A to M (corresponding to valve 1 in Figure 39)

tration the circuit remains unchanged while the valves and voltage sources interchange their positions (as in Figure 31). The progression of position changes can be followed in Figure 30 by means of the letter sequence (from A to M). (This corresponds, for example, to the conditions experienced by valve 1 of Figure 39 in one cycle.)

Complete load characteristic of the uncontrolled three-phase bridge

The direct current from the uncontrolled three-phase bridge depends upon the back-voltage E of the load. If the back-voltage is greater than the ideal direct voltage, $E \geqslant V_{dio}$, no current flows in the load. (To be more exact, in the range $V_m \geqslant E \geqslant V_{dio}$, if $L >> L_k$, a discontinuous current flows which is so small as to be negligible.) The smaller the back-voltage E, the greater is the current I_d.

This is illustrated in Figure 41, which shows sections of the direct voltage waveform $\acute{v}_d$ and the corresponding commutation of the valve current for six different levels of back-voltage E. The smaller the voltage E, the greater is the load current I_d. The diagram shows only the steady-state conditions, which develop some time after the change in the back-voltage. The steady-state condition is reached when the mean direct voltage is equal to the back voltage ($V_d = E$).

The direct voltage V_d falls with increasing current I_d because the com-

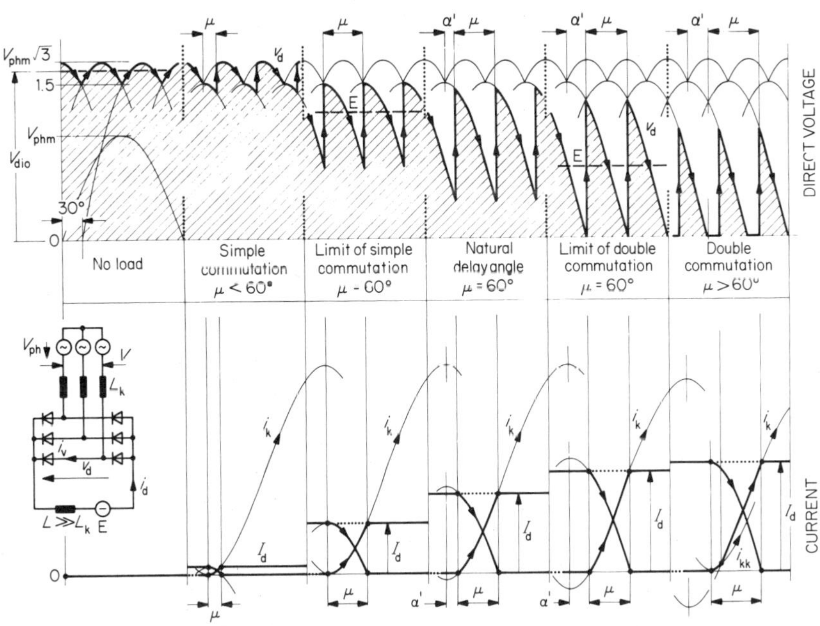

Figure 41. Typical operating conditions of the three-phase bridge for short periods in the steady state. The mean direct voltage is equal to the load back-voltage. The direct current I_d increases with decreasing back-voltage E (for graphical correlation see Figure 43)

mutation time μ, during which the variation of current produces a voltage drop, increases. The voltage–time area lost from the direct voltage applies equally to the current transfer between the commutating valves.

The direct voltage V_d falls linearly with the increasing current I_d until the overlap angle reaches a value of 60°. In this range the three-phase bridge operates according to the switching cycle of Figure 39. With increasing current I_d the duration of the single-path mode becomes progressively shorter and the commutation period longer, until, when $\mu = 60°$, one commutation is followed directly by the next. Up to this point the diodes always turn on at the point of intersection of the line voltages, since they are driven positively at that instant. However, if the direct current increases further (due to a further reduction in the back-voltage), the diodes can no longer turn on at the natural switching point, but remain on reverse, and switch only after a delay α'. The

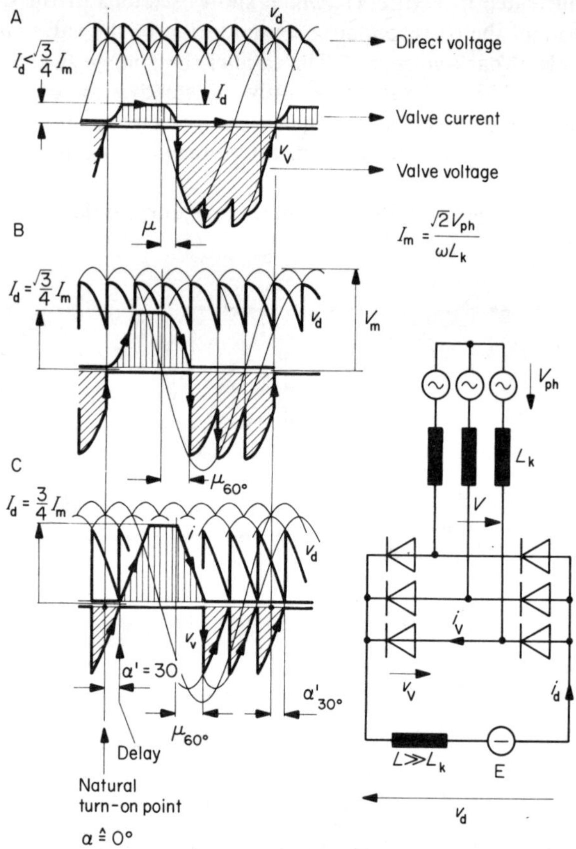

Figure 42. Direct voltage, valve voltage, and valve current in the uncontrolled three-phase bridge under three operating conditions (compare with Figure 43): (A) simple commutation: $\mu < 60°$; (B) overlap angle $\mu = 60°$: triggering at the natural switching point; (C) $\mu = 60°$: maximum shift of switching point

circuit behaves like a controlled rectifier. The direct voltage now falls with increasing current, as a result of the increasing delay α', at a steeper-than-linear rate. *The overlap angle, however, remains constant in this region*: $\mu = 60^\circ$. Figure 42 shows the valve current and voltage for the two limiting conditions of operation in the region of constant overlap $\mu = 60^\circ$, corresponding to minimum (B) and maximum (C) current. The maximum turn-on delay α' is 30°. A condition with lower load current is also shown in Figure 42 for comparison (A).

A still further increased current leads to *commutation overlap* between the positive and negative commutating groups. Accordingly, when both commutations are in progress, four valves are conducting simultaneously and the load is therefore short-circuited. The transformer is also now completely short-

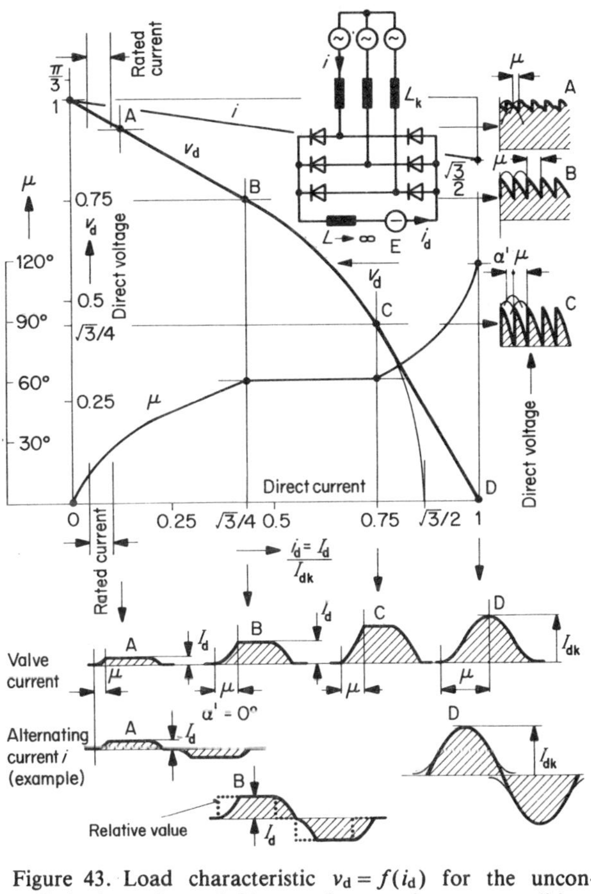

Figure 43. Load characteristic $v_d = f(i_d)$ for the uncontrolled three-phase bridge under steady-state conditions ($L \gg L_k$). Relative values. Overlap angle μ depending upon current loading. Current and voltage for arbitrary operating points (A to D; see also Figure 42). Relative alternating current i (reference value: r.m.s. value of the ideal rectilinear wave current). $v_d = E/V_{dio}$

circuited across three phases. The instantaneous direct voltage is zero (see the last period in Figure 41).

In simple commutation the commutating current i_k is determined by the line voltage V and two inductances L_k. During double commutation the current i_k is driven by the phase voltage V_{ph} through one inductance L_k. In simple commutation the tranformer experiences a single-phase short circuit and in double commutation a three-phase (limited) short circuit. In the limit $E = 0$, the transformer is completely short-circuited and loaded with a symmetrical short-circuit current. A smooth direct current I_{dk} flows in the load, with a value corresponding to the peak value I_m of the short-circuit current of the transformer.

The way in which the mean direct voltage varies with direct current can be seen in the *load diagram* $V_d = f(I_d)$ of the converter. Figure 43 shows the relationship of the direct voltage (relative to the no-load direct voltage V_{dio}) to the load current (relative to the short-circuit current I_{dk}). The overlap angle μ and the relative alternating current i are also shown in Figure 43 as functions of the direct current.

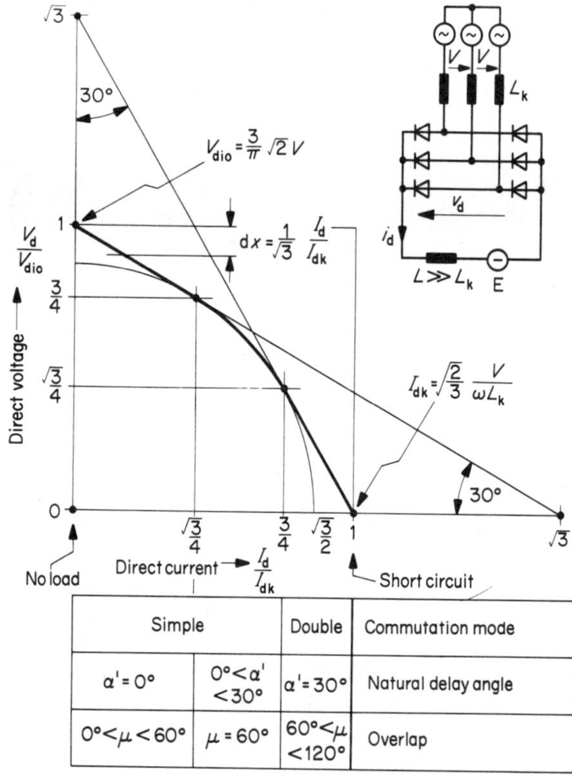

Simple		Double	Commutation mode
α' = 0°	0° < α' < 30°	α' = 30°	Natural delay angle
0° < μ < 60°	μ = 60°	60° < μ < 120°	Overlap

Figure 44. Construction of the load characteristic of the uncontrolled three-phase bridge (with smooth direct current). Summary of the three operating conditions: simple commutation, shifting of the natural switching angle α', double commutation

It will be apparent that the circuit has *three distinct operating regions.* In the first region ($0 < i_d < \sqrt{3}/4$) the direct voltage falls linearly ($1 > v_d > 0.75$) and the overlap angle increases to $\mu = 60°$. In the second region ($\sqrt{3}/4 < i_d < 3/4$) the overlap remains constant, $\mu = 60°$, and the current–voltage characteristic lies on an arc of a circle. In the last region ($3/4 < i_d < 1$) the voltage again falls linearly with increasing current up to the short-circuit condition ($E = 0$). The overlap increases from 60° to 120°.

The construction of the load diagram is elucidated in Figure 44, and the most important parameters of the three operating regions are summarized in the table.

The output characteristics in Figures 43 and 44 illustrate the whole of the possible operating range of the converter, from no-load to short circuit. Normally the rated current of the converter is from 5 to 10 per cent. of the short-circuit current, as indicated in Figure 43.

The smoothing inductor L affects only the process of adjustment when the back-voltage E changes and the degree of ripple in the direct current, not the mean value of the direct current itself, I_d (see Figure 37).

The short-circuit current I_{dk} is also independent of the load inductance L, and purely dependent upon the parameters of the transformer.

The short-circuit condition ($E = 0$) of the three-phase bridge with high load inductance ($L = 10L_k$) is illustrated in Figure 45 (from computer calculations).

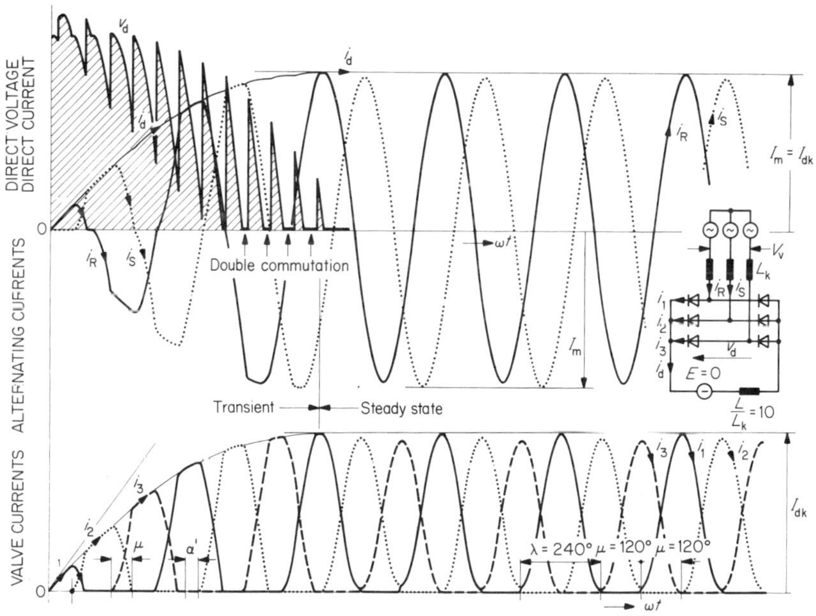

Figure 45. Switching-on process in the uncontrolled three-phase bridge. No back-voltage ($E = 0$), i.e. short circuit, point D in Figure 43. During the transient process the converter passes through all three operating conditions: simple commutation, shifting of the natural switching point (α'), and double commutation. Reproduction of a computer simulation. The apparent inequality of the steady-state currents is due to computer inaccuracy

The transient condition lasts for about one cycle. In the steady-state condition the current flowing in the transformer is theoretically a pure sine wave. The overlap is 120° and the direct voltage V_d is zero. The conduction period of the valves is 240°. The direct current I_{dk} is equal to the amplitude of the alternating current I_m. (The differences between i_1, i_2, and i_3 in Figure 45 are due to computer inaccuracy.)

Characteristics of the uncontrolled three-phase bridge

Although it is usual in the study of the three-phase bridge to neglect all ohmic resistance, and the circuit therefore contains only linear inductances and voltage sources, *the converter behaves like a d.c. voltage source V_{dio} with an internal resistance R_i.*

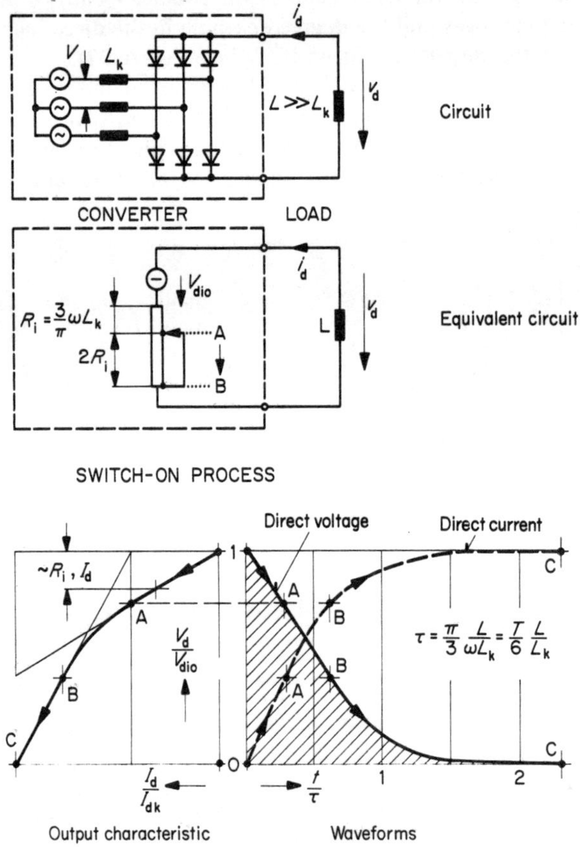

Figure 46. The uncontrolled three-phase converter behaves like a d.c. voltage source (V_{dio}) with internal resistance (R_i), in both steady-state and transient conditions. This diagram illustrates in simplified form the same process as in Figure 45. $T = 1/f$ = one cycle. $\tau = L/R$ = time constant at switch-on

In the first, linear, region of the load characteristic ($V_{dio} > V_d > \frac{3}{4}V_{dio}$) the internal resistance corresponds to the short-circuit reactance of the supply or transformer—$R_i = (3/\pi)\, X_k$. In the third region, with double commutation, the internal resistance becomes three times greater.

The switching-on transient also takes place as if in an R–L circuit. The current rises with a time constant $\tau = L/R_i$.

Figure 46 shows in a simplified form the switching-on process in a uncontrolled three-phase bridge with no back-voltage, the voltage falling as the current rises. The process follows the load characteristic through the limiting points A and B until the steady-state condition is reached at C. Between A and B the internal resistance of the converter increases by a factor of three.

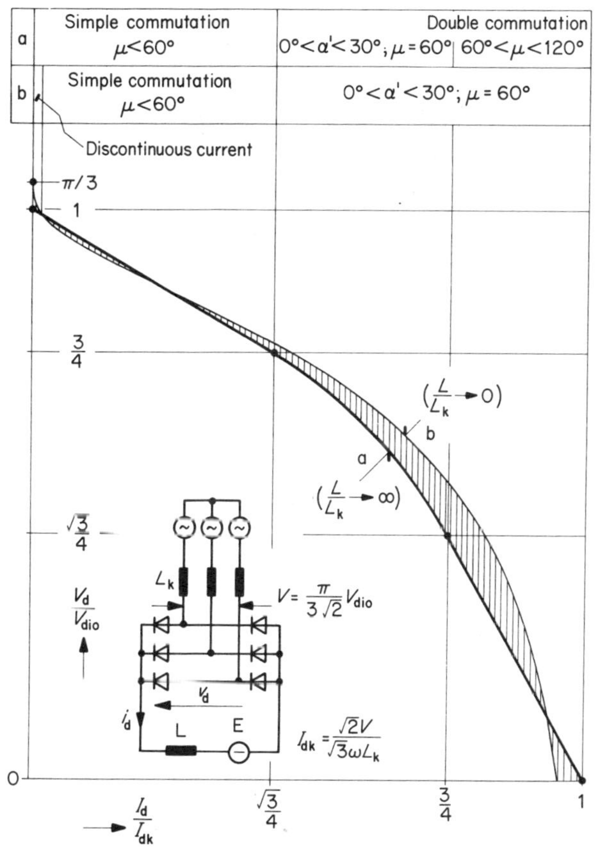

Figure 47. Load characteristics of the uncontrolled three-phase bridge for two extreme cases: very high (a) and very low (b) load inductance. With practical ratios all characteristics lie between these two extremes. The application of conventional theory (a) is therefore justified

VALIDITY LIMITS OF THE CONVENTIONAL THEORY FOR UNCONTROLLED THREE-PHASE BRIDGES

The output characteristic $V_d = f(I_d)$ of the uncontrolled three-phase bridge with L–E load is relatively independent of the inductance ratio L/L_k. Figure 47 shows the characteristics of the converter for the two extreme cases, $L/L_k = 0$ and $L/L_k = \infty$. It will be seen that the difference between the characteristics is insignificant. Without load inductance ($L = 0$) the current becomes discontinuous with a high back-voltage of $1 < E/V_{dio} < \pi/3$.

Double commutation cannot occur without inductance in the load circuit.

MATHEMATICAL APPENDIX VIII

The effective reactance (commutating reactance) of the converter transformer (Figure 33) is calculated as

$$X_k = \omega L_k = \frac{e_k V}{\sqrt{3}\, I_N} = \frac{V}{\sqrt{3}\, I_k} \tag{37}$$

where V = secondary line voltage
I_N = rated transformer current
I_k = short-circuit current with a three-phase short circuit
e_k = short-circuit voltage

During commutation the commutating current (Figure 38) is derived from the short-circuit current:

$$i_k = I'_m (1 - \cos \omega t) \tag{38}$$

where $I'_m = \dfrac{V_m}{2X_k}$

At the completion of commutation, after an interval μ, the commutating current reaches the level of the direct current:

$$i_k(\mu) = I'_m (1 - \cos \mu) = I_d$$

The overlap angle is then

$$\mu = \cos^{-1}\left(1 - \frac{I_d}{I'_m}\right) \tag{39}$$

This equation is valid only in the region of simple commutation (Figure 41), i.e. for

$$0 \leqslant \mu \leqslant 60^\circ$$

With simple commutation the direct voltage (Figure 40) is given by

$$v_d = \frac{\sqrt{3}}{2} V_m \cos \omega t \tag{40}$$

The commutating current i_k induces a voltage v_{Lk} in the reactance X_k. The resulting voltage–time integral A_x reduces the mean direct voltage by a mean (inductive) voltage drop ΔV_x to a value (see Figure 38)

$$V_d = V_{dio} - \Delta V_x \tag{41}$$

where $\Delta V_x = \frac{3}{\pi} A_x$

From the inductance law

$$v_{Lk} = \omega L_k \frac{di_k}{d\omega t}$$

from which

$$\int_0^\mu di_k = \frac{1}{X_k} \int_0^\mu v_{Lk}\, d\omega t$$

and

$$I_d = \frac{A_x}{X_k} = \frac{\pi}{3} \frac{\Delta V_x}{X_k}$$

The mean inductive voltage drop is therefore

$$\Delta V_x = \frac{3}{\pi} I_d X_k \tag{42}$$

In relative values

$$\frac{\Delta V_x}{V_{dio}} = dx$$

$$\frac{V_d}{V_{dio}} = v_d$$

$$v_d = 1 - dx \tag{43}$$

$$dx = \frac{(3/\pi) I_d X_k}{V_{dio}}$$

Further,

$$V_{dio} = \frac{3}{\pi} V_m$$

so that

$$dx = \frac{I_d X_k}{V_m}$$

From equation (37),

$$X_k = \frac{V_m}{\sqrt{3}\, I_m}$$

where $V_m = \sqrt{2}\ V$ and $I_m = \sqrt{2}\ I_k$. The relative inductive voltage drop is therefore given by

$$dx = \frac{I_d}{\sqrt{3}\ I_m} \tag{44}$$

$$I_m = \frac{V_m^{\cdot}}{\sqrt{3}\, x_k} = \frac{2}{\sqrt{3}} I'_m$$

The peak value of the three-phase short-circuit current of the transformer I_m is equal to the mean direct current I_{dk} with a short-circuit load ($E = 0$, Figures 43 and 45). As with the relative direct voltage,

$$v_d = \frac{V_d}{V_{dio}}$$

the relative direct current can be defined as

$$i_d = \frac{I_d}{I_{dk}}$$

The relative inductive voltage drop can then be expressed (see also equation (53)) as

$$d_x = \frac{1}{\sqrt{3}} i_d \qquad (45)$$

This equation again is valid only for simple commutation.

Putting $\mu = 60^\circ$ in equation (39) with a slight rearrangement gives the validity limit for the current

$$i \leqslant \frac{\sqrt{3}}{4}$$

This gives in addition $dx \leqslant \frac{1}{4}$ and $v_d \geqslant \frac{3}{4}$.

THE CONTROLLED THREE-PHASE BRIDGE

Converter control—general

One of the most important converter circuits in power electronics is the controlled three-phase bridge, using *thyristors* as the valves (Figures 4 to 7). As well as the protection and cooling devices with which the thyristors must be provided, an essential part of the converter is a *control system*. The control system produces the triggering pulses necessary for the thyristors. These *control pulses* are then conveyed to the *gates* of the thyristors by means of *pulse couplers*. So long as the thyristor is supporting positive voltage and the current path is completed, the thyristor can be triggered by the control pulse and will turn on.

The control pulses are *synchronized* to the supply voltage. *In each period, the thyristors receive two control pulses, separated by* 60°. The reference point for the timing of the first pulses (triggering angle α) is the natural turning-on point that would be experienced by diodes (see Figures 29 and 38). Thus operation with a triggering angle $\alpha = 0^\circ$ corresponds to that of an uncontrolled diode bridge.

The triggering pulses can be phase-shifted, relative to the reference point, in the control system by a control voltage. (For the case of a single-pulse converter, see Figure 12.) In this way the turning-on instant of the thyristors can be varied, with a delay which is designated the control angle, or delay angle α.

The phase-shifting of the triggering pulses—the variation of the triggering angle α—can readily be effected by an external control signal. The effect of shifting the triggering pulses, as will be explained further, is to vary the output voltage of the converter. In a power control system, the converter functions as a regulator or an amplifier. The control characteristic and the dynamic behaviour of the converter are important factors in control.

In comparison with the converter circuit, the circuit of the control system is quite complicated. The generation, matching, and phase-shifting of the pulses can be achieved by means of a wide variety of techniques. In small converters the control is combined with the power circuit in a single unit. In a large converter installation the control units are mounted separately.

Details of control systems do not form a part of converter circuit theory. It is simply accepted that the thyristors are turned on by triggering pulses at any desired delay angle α. For simplicity, the control units are also omitted from the converter diagrams.

Triggering pattern for the three-phase bridge

In the absence of triggering pulses the thyristors are non-conducting and, in the steady state, support in the three-phase bridge the phase voltage of the

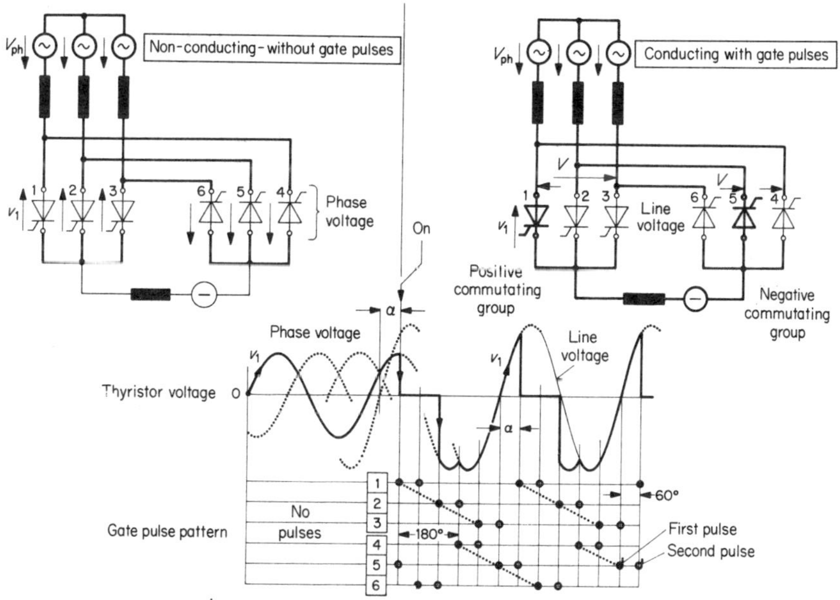

Figure 48. Controlled three-phase bridge. Without triggering pulses all the thyristors are blocking and support the phase voltage (left). With triggering pulses (arbitrary value of α) and with load the line voltage is applied to the thyristors. At switch-on thyristors 1 and 5 receive pulses simultaneously

supply. When triggering pulses are applied to the thyristors and the converter is loaded, the valves that are not conducting are subjected to the transformer line voltage. This is shown graphically in Figure 48.

In the first instance *two thyristors must be triggered simultaneously* (in Figure 48, for example, 1 and 5). For this reason, two pulses, 60° apart, are applied to each thyristor in a cycle. (Pulses are assumed to be very narrow.)

In each commutating group the thyristors are triggered in sequence at intervals of 120°. The control pulse sequences of the two groups are mutually displaced by 180°. The pulse sequence of the two groups are mutually displaced by 180°. The pulse sequence is illustrated in Figures 48 and 49. These show all the thyristor voltages for an arbitrary delay angle ($\alpha = 60°$: commutation is neglected, $\mu = 0$).

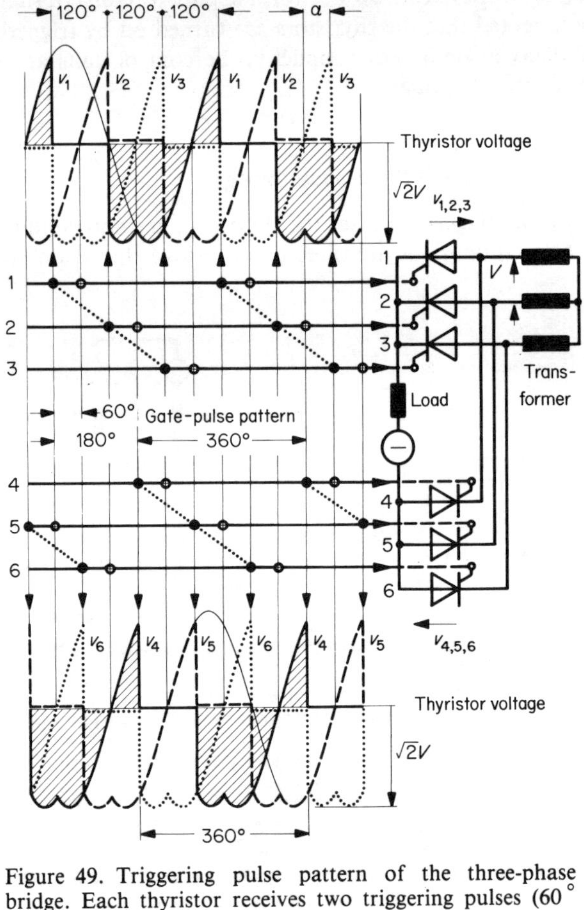

Figure 49. Triggering pulse pattern of the three-phase bridge. Each thyristor receives two triggering pulses (60° apart) per cycle. In each commutating group (1–2–3 and 4–5–6) the thyristors are triggered periodically at intervals of 120°. The triggering pulses of the two commutating groups are mutually displaced by 180° (commutation neglected)

At least two thyristors of the three-phase bridge are always conducting simultaneously. The double-gate pulse is required for the initial triggering, one thyristor being triggered by its first pulse, the other by its second. In steady-state operation, however, with continuous current, the second pulse is not necessary. On the other hand, discontinuous current operation can only be maintained by means of the double pulse.

Control of direct voltage—ideal conditions

The converter functions as a d.c. voltage source. The principal converter parameter is the mean value of the direct voltage V_d, measured at the d.c. terminals. The maximum mean direct voltage of the controlled three-phase bridge is obtained on no-load in the full-output condition ($\alpha = 0^\circ$), corresponding to diode rectifier operation (see Figures 27 and 29). This value is designated the *ideal no-load direct voltage* V_{dio} and is simply related to the transformer voltage V:

$$V_{dio} = \frac{3}{\pi}\sqrt{2}\,V \qquad \text{(see equations (30) and (31))}$$

The converter cannot supply a higher voltage.

A reduction of the direct voltage V_d can be effected steplessly by the control. This is illustrated in Figure 50. The voltage and current waveforms are

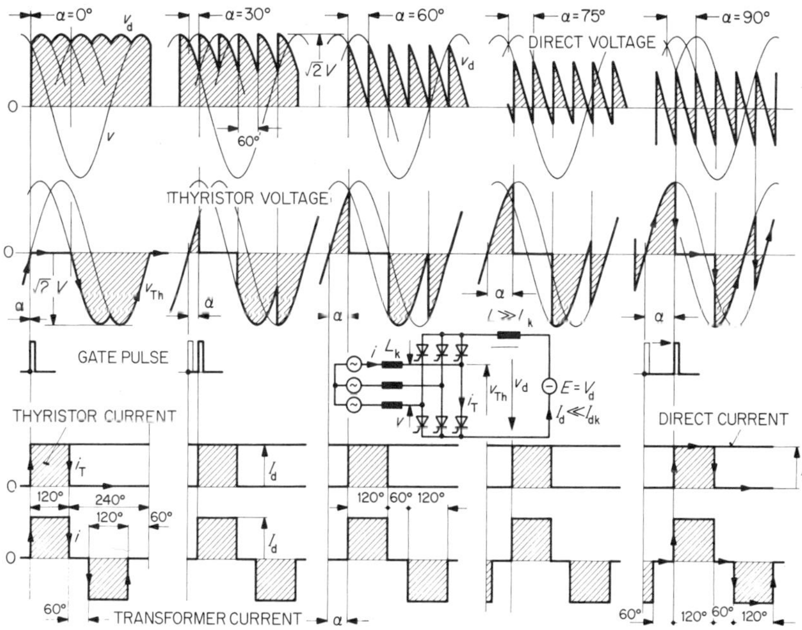

Figure 50. Controlled three-phase bridge. Voltage and current with different delay angles α. Idealized representation with commutation neglected. The direct voltage falls with increasing delay angle α (for the control characteristic see Figure 51). Steady-state conditions with constant load current

idealized here and commutation is neglected. (It is assumed that $L/L_k \to \infty$, $I_d = \text{constant}$, and $\mu \to 0$.) (see Figure 70)

The direct current I_d in this case is completely smooth, commutation is instantaneous and the currents in the thyristors and the transformer exhibit rectilinear waveforms.

The output condition $\alpha = 0°$ corresponds to the operation of the uncontrolled converter (with diodes). The direct voltage is formed of the envelope of the positive and negative line voltages (the mean value V_{dio} is only about 5 per cent. less than the amplitudes). The maximum voltage to which the valves are subjected in the reverse direction is the peak value of the transformer voltage $\sqrt{2}\ V$. The conduction period of the thyristors is 120° (one-third cycle) and the amplitude of the thyristors current is equal to the load current I_d. The transformer current i is composed of two consecutive valve current pulses.

The control pulses (the figure shows the first pulse) are now shifted to the right. This means that the thyristors are triggered with a delay α relative to the natural switching instant. During the delay period α the direct voltage follows the falling transformer voltage. At the instant of triggering and instantaneous commutation the direct voltage jumps to the next voltage sine wave, and the same process is repeated six times per cycle. *As a result of the control—the triggering delay—the mean value of the direct voltage is reduced.* When $\alpha = 90°$ the mean direct voltage is equal to zero. It will be seen in Figure 50 (right) that in this case the direct voltage is actually an alternating voltage, of six times the supply frequency. The thyristor is then subjected equally to positive and negative voltages. In all of the five control conditions in Figure 50 the direct current is constant (i.e. the back-voltage E is in all cases matched to the direct voltage V_d) and only the steady-state conditions are shown.

MATHEMATICAL APPENDIX IX

The mean value of the ideal direct voltage (without commutation) of the controlled three-phase bridge (Figure 50) is calculated as follows:

$$V_{di} = \frac{3\sqrt{2}\ V}{\pi} \int_{60°+\alpha}^{120°+\alpha} \sin \omega t \ d\omega t$$

$$= \frac{3}{\pi} \sqrt{2}\, V \cos \alpha$$

Substituting from equation 30,

$$V \frac{3}{\pi} \sqrt{2} = V_{dio}$$

$$V_{di} = V_{dio} \cos \alpha \tag{46}$$

In relative terms

$$v_{di} = \frac{V_{di}}{V_{dio}} = \cos \alpha \tag{47}$$

Figure 51 shows the graphical representation of equation (47). Also plotted are the relative r.m.s. values of the direct voltage and the thyristor voltage.

Under all steady-state conditions

$$E = V_{di}$$

For the rectilinear wave thyristor current:

$$\text{Mean current: } I_{Thav} = \frac{1}{3} I_d \tag{48}$$

$$\text{R.M.S. current: } I_{Thrms} = \frac{1}{\sqrt{3}} I_d \tag{49}$$

The r.m.s. value of the (rectilinear wave) transformer current is given by

$$I = \frac{\sqrt{2}}{\sqrt{3}} I_d \tag{50}$$

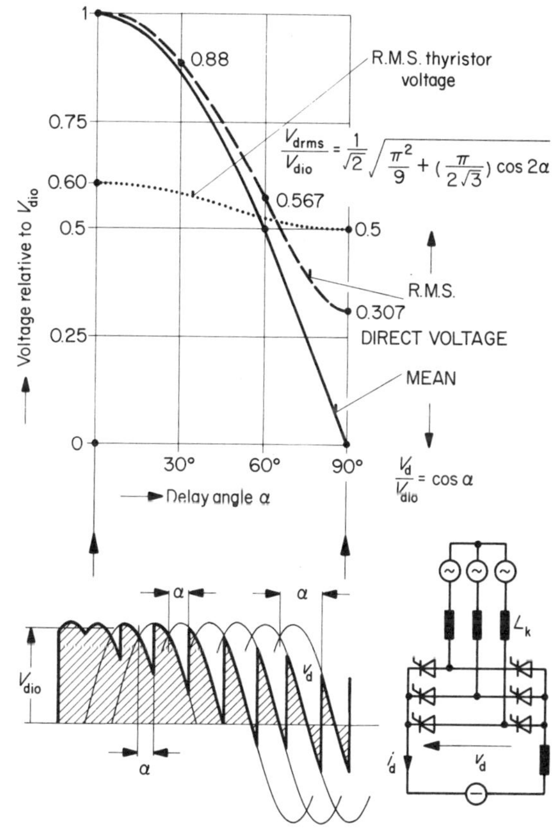

Figure 51. Control characteristic of the three-phase bridge. Idealized representation (instantaneous commutation). The diagram corresponds to the waveforms of Figure 50. Steady-state conditions with constant load current

The r.m.s. transformer current is $\sqrt{2}$ times greater than the r.m.s. thyristor current, since it consists of two valve current pulses.

OUTPUT CHARACTERISTIC OF THE CONTROLLED THREE-PHASE BRIDGE IN RECTIFIER MODE

Maximum-output condition, $\alpha = 0^\circ$

In considering the uncontrolled three-phase bridge (Figures 41 and 42) it was shown that the direct voltage V_d of the converter falls with increasing current loading I_d. The converter behaves like a d.c. voltage source V_{dio} with an internal resistance R_i (Figure 46), although no resistances are included in the circuit. Physically, the origin of this regulation of the converter lies in the lost voltage–time integral during commutation (Figure 38).

Three distinct operating regions can be observed in the load diagram of the uncontrolled three-phase bridge with smooth direct current (Figures 43 and 44): the regions of simple commutation, constant overlap, and double commutation. This applies, however, only on the assumption of a very high load inductance, $L/L_k \to \infty$, and under steady-state conditions. The fully controlled three-phase bridge, embodying thyristors, with a delay angle $\alpha = 0^\circ$ (the natural point of commutation) exhibits the same behaviour in the region of simple commutation ($\mu < 60^\circ$) as the uncontrolled bridge with diodes. In the region of constant overlap angle ($\mu = 60^\circ$), however, the natural point of

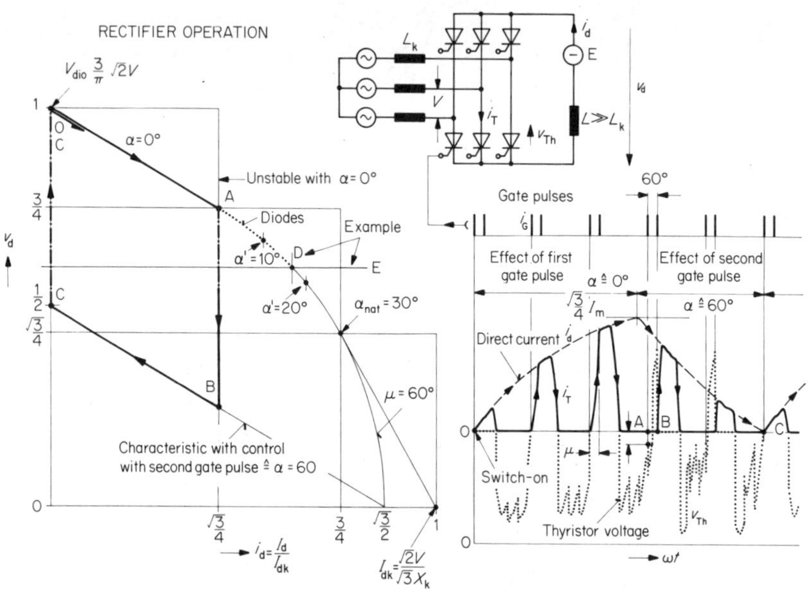

Figure 52. Load characteristic of the controlled three-phase bridge with constant delay angle $\alpha = 0^\circ$. With $E/V_{dio} < \frac{3}{4}$ a stable condition on the 'diode characteristic' cannot be achieved. Example: $E/V_{dio} = 0.6$, $L/L_k = 10$. The intended operating point D cannot be obtained: the load current oscillates. Since the rated current is generally well short of this point the effect is not normally very significant

commutation shifts gradually from 0 to 30° and the triggering pulse has therefore to be delayed. If the triggering pulse remains in the position $\alpha = 0°$, the thyristor cannot be triggered by the first pulse and turns on only as a result of the second pulse. Since the two triggering pulses are 60° apart (Figure 49), the converter switches abruptly into an operating mode corresponding to a delay angle $\alpha = 60°$. The converter has no stable operating condition.

The *unstable behaviour of the three-phase bridge* in the region $0° \leqslant \alpha' \leqslant 30°$ and $E/V_{dio} \leqslant \frac{3}{4}$ (i.e. for $I_d/I_{dk} > \sqrt{3}/4$) is illustrated in Figure 52. The example ($L/L_k = 10$, $E/V_{dio} = 0.6$) shows that the desired operating point (D) in the maximum-output condition of the converter ($\alpha = 0°$) is not attainable. Following the switching-on (point O) the current rises in accordance with the characteristic for $\alpha = 0°$ (as for the diode bridge). At the operating point A ($V_d/V_{dio} = \frac{3}{4}$, $I_d/I_{dk} = \sqrt{3}/4$) the thyristors cannot be triggered, because the voltage applied to them at the occurrence of the pulse is still negative. Only at point B, with the second pulse, is the thyristor voltage positive and the thyristor capable of being triggered. Triggering continues to occur with the second pulse, corresponding to a delay angle of 60°. Since the back-voltage E is higher than the maximum value of the direct voltage with $\alpha = 60°$, the current falls to zero (point C). The original situation is thereby restored and the whole process is repeated. The converter is placed in an unstable condition, resulting in oscillations of the direct current. The critical point occurs at $\mu = 60°$.

The converter with zero delay is not stable in the range $\frac{1}{2} < E/V_{dio} < \frac{3}{4}$. With a higher back-voltage, $\frac{3}{4} < E/V_{dio} < 1$, the characteristic of the thyristor bridge ($\alpha = 0°$) is identical with that of the diode bridge. With lower back-voltages, $0 < E/V_{dio} < \frac{1}{2}$, the steady-state characteristic of the bridge with zero delay is that corresponding to $\alpha = 60°$ (triggered by the second gate pulse, displaced by 60°). This is illustrated in Figure 53, which shows, for five different values of the back-voltage E, the transient process in the converter from switch-on to the attainment of the steady-state characteristic.

The current loading of the three-phase bridge converter is usually (in d.c. motor drives, for example) considerably below the critical value $I_d/I_{dk} = \sqrt{3}/4$. Thus in normal operating circumstances the unstable condition described does not arise.

Control range $0° \leqslant \alpha \leqslant 90°$

By phase-shifting the control pulses the delay angle α of the converter is varied, which, as shown in a simplified way in Figure 50, varies the direct voltage v_d. In the ideal case the mean value of the direct voltage is given by $V_d = V_{dio} \cos\alpha$. The maximum obtainable peak value of the direct voltage is approximately equal to the amplitude of the line voltage $V_{dio} = (3/\pi)V\sqrt{2}$. With $\alpha = 90°$ the mean value V_d is zero, but the terminal voltage of the converter is not zero under this condition (Figure 50, top right), but is an alternating voltage of triangular waveform.

Commutation, the transfer of current from one valve to another, causes a reduction of the direct voltage. Figure 38 shows the voltage–time integral (A_x) lost through commutation in the case of the uncontrolled converter. The mean

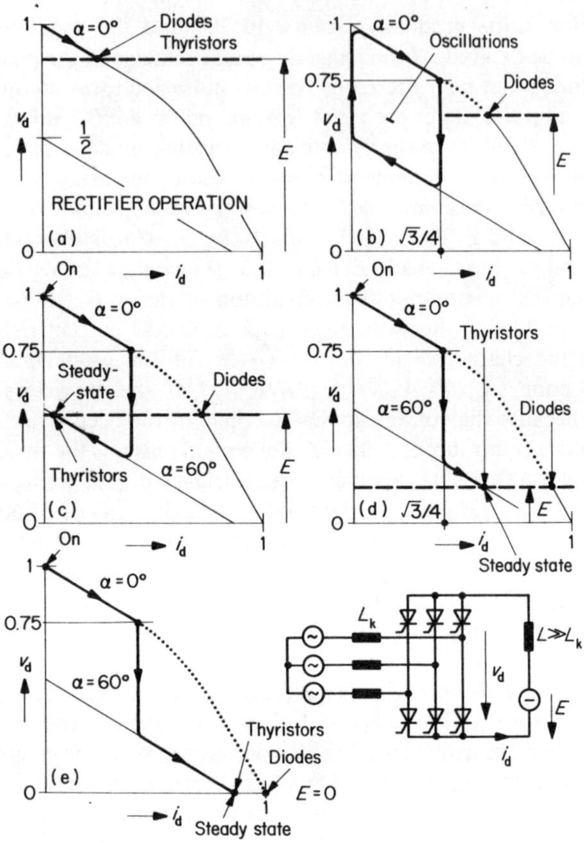

Figure 53. Principal operating conditions of the three-phase bridge with minimum delay ($\alpha = 0^\circ$)(narrow triggering pulse). (a) $E/V_{dio} > \frac{3}{4}$: the same characteristic for the diode and thyristor bridges; (b) $\frac{1}{2} < E/V_{dio} < \frac{3}{4}$: oscillations (see Figure 52); (c), (d), and (e) $0 < E/V_{dio} < \frac{1}{2}$: steady-state on the 60° characteristic

voltage drop, with simple commutation, is proportional to the direct current: $\Delta V_x = (3/\pi)X_k I_d$.

This rule applies also to the controlled converter. With constant current loading, therefore, the voltage drop ΔV_x is independent of the delay angle α. With a constant triggering angle, for any value of α, the direct voltage V_d falls linearly with increasing load current. *The load characteristics of the controlled converter are parallel to one another, with* α as the parameter.

Figure 54 shows how the direct voltage v_d varies with the delay angle for a constant direct current I_d. It will be observed that *with an increasing delay angle the overlap angle decreases, but that the lost voltage-time integral A remains constant.* With $\alpha = 0^\circ$ the commutation current i_k follows its waveform at around 90°, the rate of change of current is small, and the commutation takes a relatively long time. With $\alpha = 90^\circ$ the commutating current is in the middle of the sinusoidal short-circuit current waveform, the rate

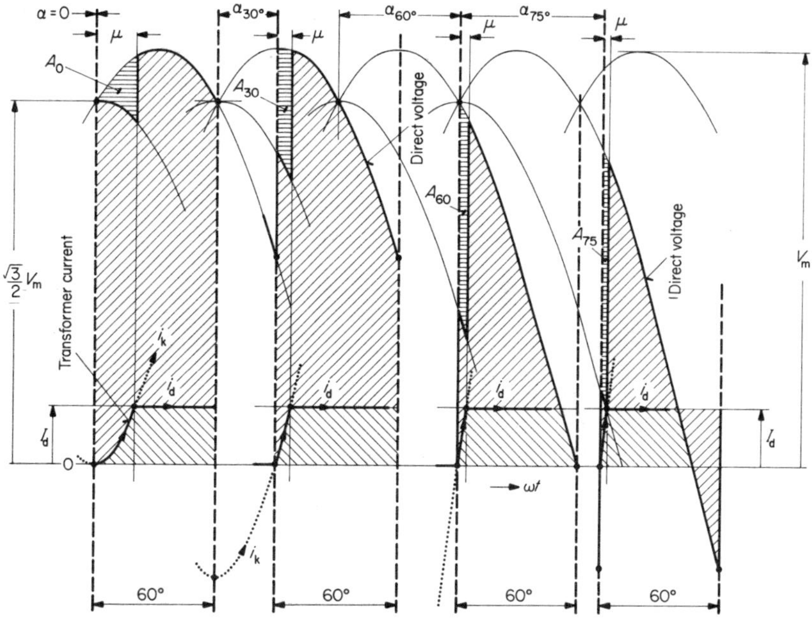

Figure 54. Direct voltage of the three-phase bridge with various delay angles (sections). With constant current loading the voltage–time areas (A) lost through commutation are the same in all cases. The overlap angle μ decreases with increasing delay angle α

of change of current, di/dt, is at its highest, and the commutation period is shorter.

During commutation (μ) the direct voltage follows a sine wave with an amplitude $(\sqrt{3}/\sqrt{2})\,V$ and a phase shift of 30° with respect to the line voltage V. For both commutation and the single-path mode of operation the same equivalent circuit may be employed for the controlled converter as for the uncontrolled bridge (Figure 40).

The commutations are reflected both in the direct voltage and in the thyristor voltage. The beginning and end of each commutation can be discerned in the thyristor voltage (Figure 55). Both positive and negative voltage steps are presented to the thyristors as a result of commutation. The conduction period of the thyristors is

$$\lambda = 120^\circ + \mu$$

The load diagram of the controlled three-phase bridge is shown in Figure 56. It relates to the steady-state conditions of the converter and for an inductance ratio $L/L_k \to \infty$. The converter is here functioning as a rectifier over the whole range. The maximum direct current I_{dk} occurs with $E = 0$ and $\alpha = 30^\circ$. With $\alpha = 0^\circ$ the short-circuit current lies on the 60° characteristic, and is therefore less than with $\alpha = 30^\circ$. *The maximum direct current I_{dk} is equal to the amplitude of the symmetrical short-circuit current of the transformer, I_m.*

The load characteristics of the controlled three-phase bridge are linear so long as the overlap angle μ is less than 60°. The critical value $\mu = 60^\circ$ lies on

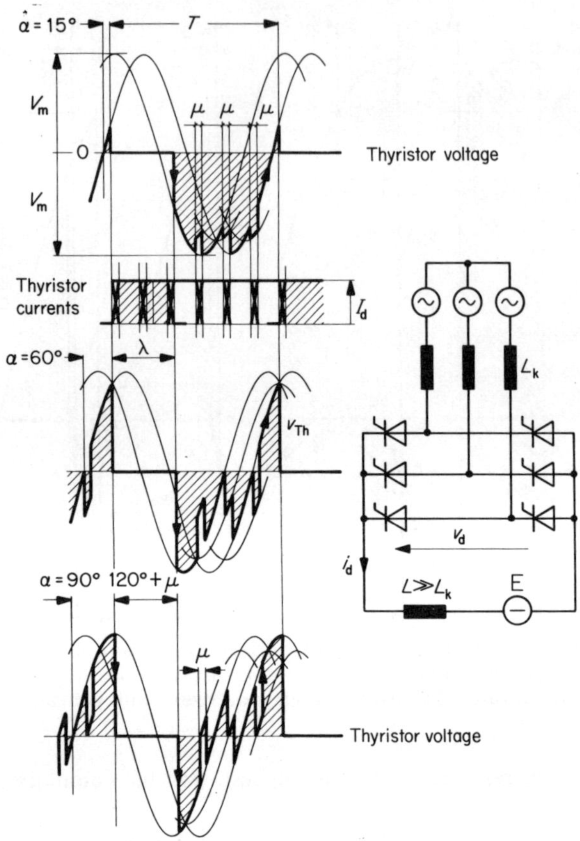

Figure 55. Thyristor voltage in the controlled three-phase bridge with three different delay angles. Operation with simple commutation. Steady-state conditions. Six commutations take place in one cycle; three of these are evident in the thyristor voltage

a circle in the relative load diagram. Overstepping this limit leads to instability phenomena in the range $\sqrt{3}/4 < v_d < \frac{3}{4}$ (Figures 52 and 53) and to double commutation in the range $0 < v_d < \sqrt{3}/4$.

Figure 57 illustrates the current and voltage waveforms in the three-phase bridge. A delay angle α of 30° is taken as an example and the load current I_d is assumed to be smooth. The most important formulae for the three-phase bridge are summarized in Figure 58.

The converter attains a steady-state condition when the mean direct voltage V_d is equal to the back-voltage E. With a constant delay angle α and a constant back-voltage E, the voltage V_d, after switching on, falls from its initial value V_{di} until the condition $V_d = E$ is established. The direct current I_d can only flow if the output voltage V_{di} is higher than E. The greater the difference ($V_{di} - E$), the higher is the steady-state current. *The direct current I_d from the converter depends upon the delay angle α and the back-voltage E.* The relationships are illustrated in Figure 59. If the back-voltage E is higher than the no-load voltage V_{dio}, no current can flow in the converter (a). With no back-voltage ($E = 0$)

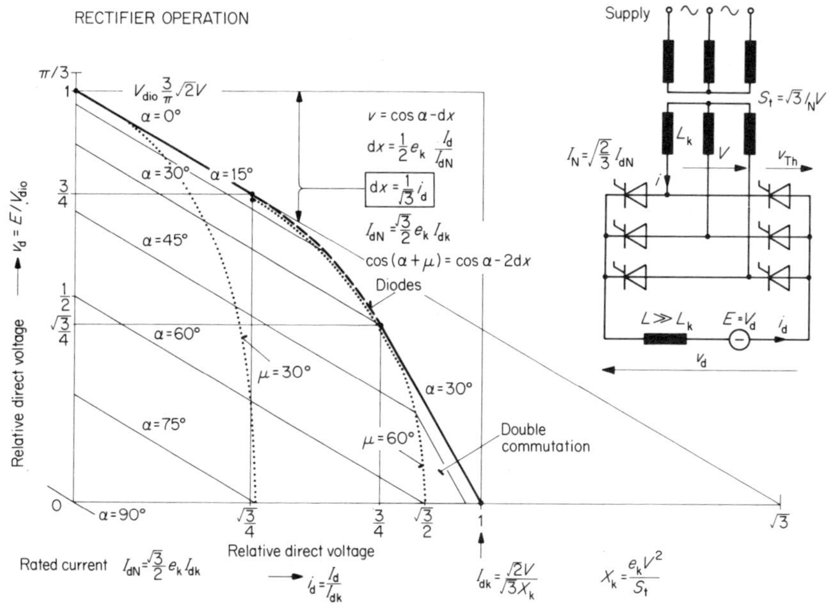

Figure 56. Load characteristics of the controlled three-phase bridge. Steady-state conditions with smooth direct current. Current i_d in relative values. Up to an overlap $\mu = 60°$ all the characteristics, with the delay angle α as parameter, are parallel. The effect of the supply is neglected. L_k = transformer short-circuit inductance. S_t = transformer power rating

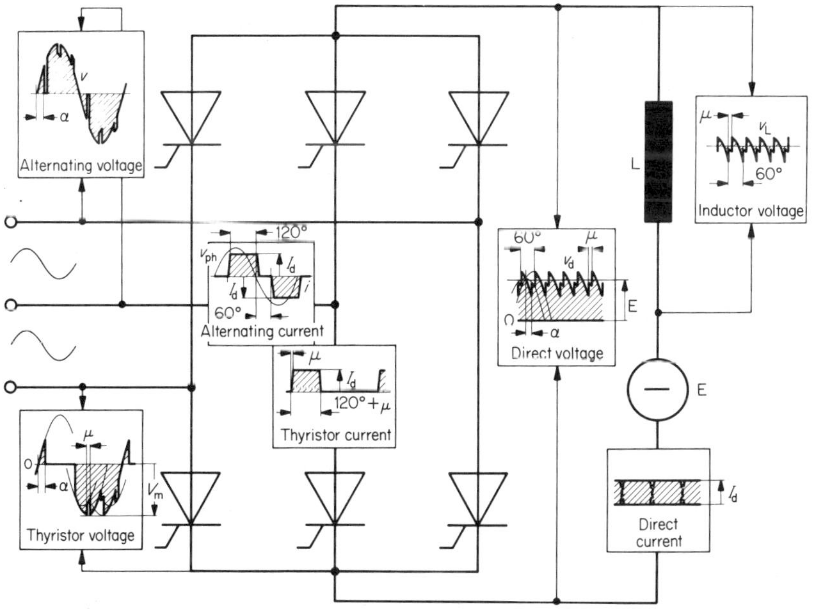

Figure 57. Typical current and voltage waveforms in the controlled three-phase bridge assuming smooth direct current. An arbitrary delay angle $\alpha = 30°$ taken as an example. Steady-state condition. Operation with simple commutation. The delay angle particularly affects the waveforms of the thyristor voltage and the direct voltage

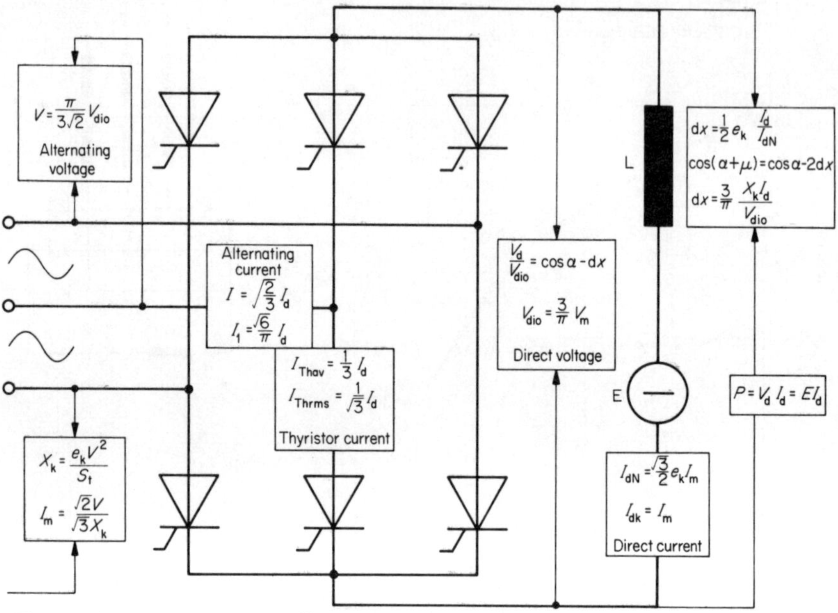

Figure 58. Mathematical relationships in the controlled three-phase bridge according to conventional theory (smooth direct current). The formulae are valid only for the range where $\mu < 60°$ (see load diagram, Figure 56). Commutation neglected in respect of alternating current and voltage

the current attains the short-circuit value (b). If there is a back-voltage, for the same delay angle, the current becomes smaller (e). The same direct current can be obtained with different delay angles and back-voltages (c and e). With a given back-voltage, a different direct current can be obtained with a different delay angle (c and d).

MATHEMATICAL APPENDIX X

By substitution from equations (41) and (46), the direct voltage from the controlled three-phase bridge, with allowance for simple commutation, can be calculated as follows:

$$V_d = V_{dio} \cos\alpha - \Delta V_x = V_{dio}\left(\cos\alpha - \frac{\Delta V_x}{V_{dio}}\right) \tag{51}$$

where

$$V_{dio} = \frac{3}{\pi} V_m \qquad \text{(ideal no-load direct voltage)}$$

and from equation (42),

$$\Delta V_x = \frac{3}{\pi} I_d X_k \qquad \text{(inductive voltage drop)}$$

In relative terms (with respect to V_{dio})

$$v_d = \cos\alpha - dx \tag{52}$$

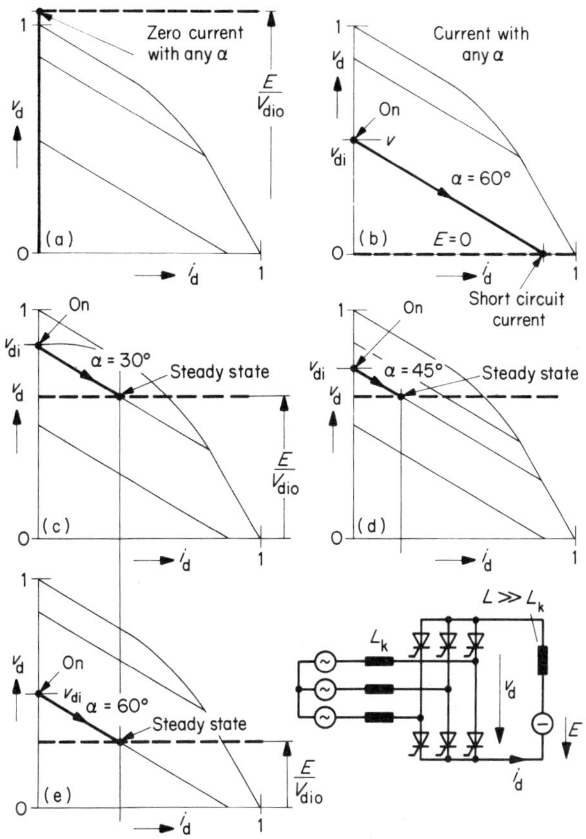

Figure 59. Load characteristic area of the controlled three-phase bridge with smooth direct current. Selected examples of steady-state operating points with various delay angles α and back-voltages E

where $\quad \mathrm{d}x = \dfrac{\Delta V_x}{V_{\mathrm{dio}}}$

The relative inductive voltage drop dx is independent of the delay angle, and therefore equal to that in the uncontrolled converter (equation (44)). Substituting from equations (37) and (50),

$$\mathrm{d}x = \frac{1}{2} e_k \frac{I_d}{I_{dn}} \tag{53}$$

The rated converter current is here

$$I_{dn} = \sqrt{\frac{3}{2}} I_N = \frac{\sqrt{3}}{2} e_k I_{dk}$$

$$I_{dk} = I_m = \frac{\sqrt{2}V}{\sqrt{3}X_k}$$

$$X_k = \frac{e_k V^2}{S_t}$$

$$S_t = \sqrt{3}\, I_N V \qquad \text{(rated apparent power of the transformer)}$$

By analogy with equation (38), the commutating current is given, according to equation (3), by

$$i_k = I'_m\,[\cos\alpha - \cos(\omega t + \alpha)] \tag{54}$$

where $I'_m = \dfrac{V_m}{2X_k}$

During the period $\omega t = \mu$ the current i_k reaches the level of the direct current I_d, and therefore

$$I_d = I'_m\,[\cos\alpha - \cos(\mu + \alpha)]$$

or

$$\cos(\mu + \alpha) = \cos\alpha - \frac{I_d}{I'_m} \tag{55}$$

Then, since

$$I'_m = \frac{\sqrt{3}}{2} I_{dk}$$

from equations (53) and (44),

$$\cos(\mu + \alpha) = \cos\alpha - 2dx \tag{56}$$

This establishes the relationship of the overlap angle μ to the delay angle α and the current loading. All the equations given above are valid only so long as $\mu < 60°$.

These equations are repeated in Figure 56.

In all cases, under steady-state conditions, $E = V_d$.

LOAD CHARACTERISTIC OF THE CONTROLLED THREE-PHASE BRIDGE IN THE INVERTER MODE

General

The controlled converter can fulfil two different functions: either rectification of an alternating voltage or the inversion of a direct voltage. In the simplest arrangement, based on the example of a single-pulse converter, the operation of the converter as a rectifier or as an inverter was illustrated in Figure 23. It is established that:

In a rectifier energy is transmitted from the a.c. voltage source to the d.c. voltage source. In an inverter energy is supplied in the reverse direction from the d.c. voltage source to the a.c. voltage source.

To reverse the direction of energy flow, the polarity of the d.c. voltage source must be reversed and the inverter suitably controlled. The direction of current flow is not changed.

In the rectifier mode the thyristors support predominantly negative voltage; in the inverter mode, positive voltage.

The turning-off of the thyristors—commutation—is in both cases accomplished by the a.c. voltage source—the supply. In both rectifier and inverter operation, the converter is *supply-commmutating;* the commutation can be regarded as 'natural'.

Without an alternating voltage supply, the inverter with natural commutation (in contrast to the forced-commutation inverter) cannot produce an a.c. output. The bidirectional flow of energy can thus take place only between two actual voltage sources (a.c. and d.c.).

The dual function of the converter is clearly seen in d.c. motor drives. *In the rectifier mode the converter drives the machine as a motor. When the machine functions as a generator (during braking, for example) the converter functions as an inverter.*

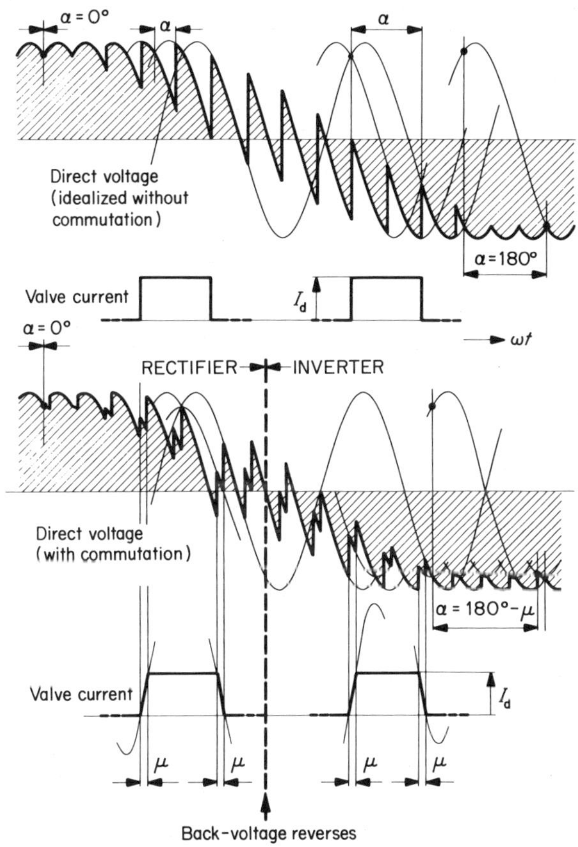

Figure 60. Waveform of direct voltage in the three-phase bridge during the transition from rectifier to inverter operation. Diagrammatic representation assuming constant smooth direct current. The delay angle is shifted from $\alpha = 0°$ to $\alpha = 180°$. In practice the transition usually entails a period of discontinuous current

Inverter characteristics

In the steady-state condition of the converter the mean direct voltage V_d corresponds to the back-voltage E of the load. With a variable back-voltage the direct current can be held constant by suitable control of the delay angle (Figure 59 c and e). This applies not only in the case of positive direct voltage, but also with negative voltage. This is shown diagrammatically in Figure 60. With the direct current held constant, the direct voltage is changed from a positive to a negative direction by the control. In the process, the back-voltage E must always be varied to match the mean direct voltage V_d. It can be seen that the voltage can reverse with the direction of current unchanged. Accordingly the power P, being the product of direct current and direct voltage, changes its sign. The converter functions as an inverter. The flow of power is directed from the d.c. voltage source to the a.c. voltage source.

In the ideal case (no commutation—smooth direct current, Figure 60, top)

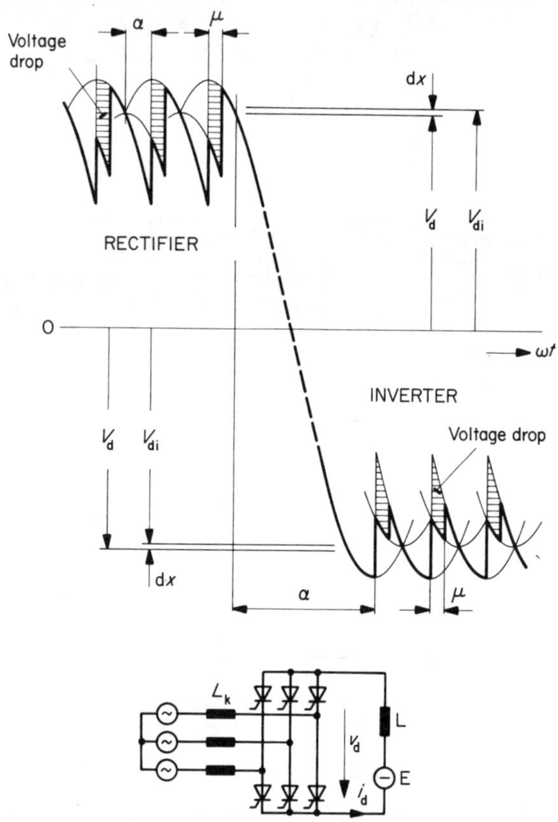

Figure 61. The inductive direct voltage drop dx due to commutation with constant direct current is the same in the rectifier and inverter modes. The (negative) inverter voltage, unlike that of the rectifier, is increased by the voltage drop

the same maximum direct voltage can be obtained in the negative as in the positive direction. In fact, when the commutation voltage drop is taken into account, it is still the case that *the same maximum direct voltage (negative) is attained in the inverter as in the rectifier.*

In the rectifier operating mode the maximum direct voltage is produced, for a given delay angle α, with minimum load current ($I_d \to 0$). With increasing current the voltage falls. *In the inverter operating condition, in contrast, the (negative) direct voltage increases with increasing current loading.* The (negative) voltage drop dx is added to the (negative) direct voltage.

This is shown in Figure 61. With a given direct current the voltage drop dx is the same for both rectification and inversion. The positive direct voltage V_{di}, however, is reduced by the inductive voltage drop dx, whereas the negative direct voltage is increased.

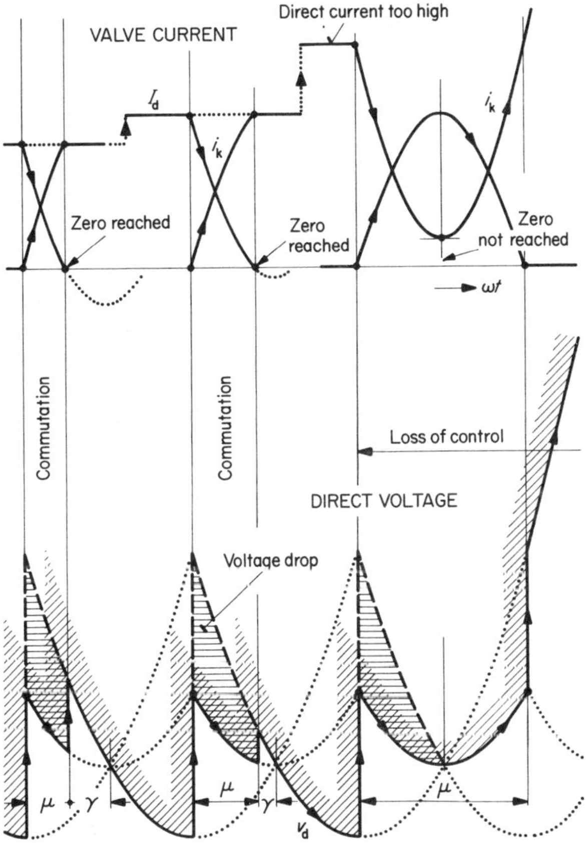

Figure 62. Commutation in the inverter mode is successful so long as $\alpha + \mu < 180°$. If this limit is exceeded, the inverter loses control. Diagrammatic representation of three commutation processes with constant delay angle $\alpha = 150°$ and increasing current, to the point of losing control. μ = overlap angle; γ = safety margin

As the current rises, the inverter voltage cannot increase indefinitely. It can be observed in Figure 62 that, initially, the inductive voltage drop increases with the direct current, and so, therefore, does the mean direct voltage. The voltage drop is caused by the commutation, and is proportional to the direct current I_d. During commutation, one valve takes over the current from another. Commutation is complete when the first conducting thyristor turns off. This can only happen when the commutation current passes through zero.

Figure 62 shows that beyond a certain current limit the commutating current i_k does not fall to zero, so that the current cannot be transferred to the next valve. The first thyristor continues to conduct and commutation is not achieved.

The inverter lapses into a short-circuit condition and is no longer controllable (commutation failure); the direct current increases continuously.

It can be seen in Figure 62 that the limit of inverter control is reached when $\mu + \alpha = 180°$. The overlap angle μ is a function of the direct current I_d and the delay angle α. For any particular angle α there is a corresponding value of the direct current I_d which satisfies the condition $\mu + \alpha = 180°$.

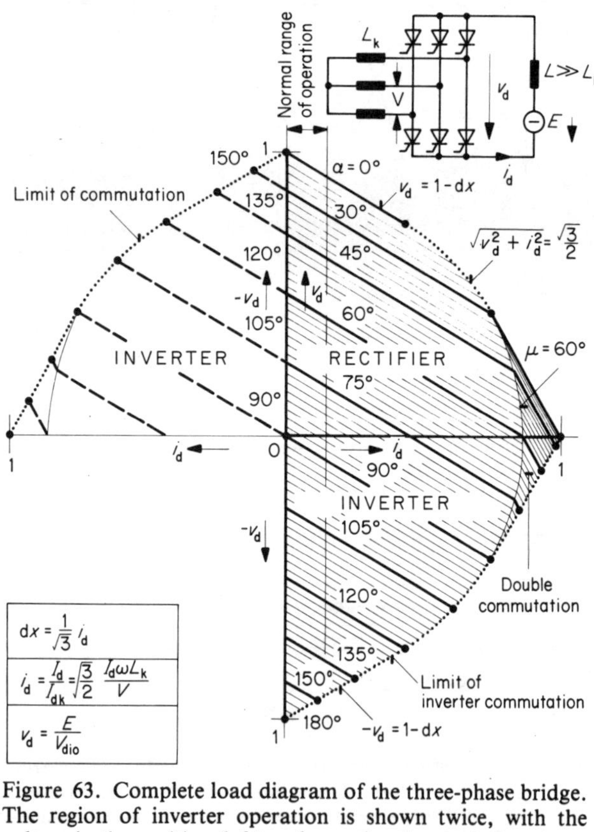

Figure 63. Complete load diagram of the three-phase bridge. The region of inverter operation is shown twice, with the voltage in the positive (left) and negative (bottom) directions. Delay angle α as parameter. Conventional representation with smooth direct current under steady-state conditions

The operating area of the inverter is bounded by the limit of control. This limit is symmetrically disposed with regard to the limiting rectifier characteristic. That is to say, *at a given current the converter can develop the same maximum direct voltage whether as a rectifier or as an inverter.*

Figure 63 illustrates the load diagram of the converter for rectifier and inverter operation. The characteristics, with the delay angle α as parameter, in part run smoothly from rectifier into inverter operation. Again, the diagram is accurate only for steady-state (or slowly varying) conditions and for very high load inductances. The inverter characteristics appear twice in Figure 63. Beneath the rectifier diagram the characteristics run as a logical extension into the region of negative voltage. On the left the inverter voltage is considered as positive, illustrating clearly the fact that the inverter voltage, unlike the rectifier voltage, increases with increasing current loading.

MATHEMATICAL APPENDIX XI

The direct voltage equation (52),

$$v_{\mathrm{d}} = \cos\alpha - \mathrm{d}x \tag{57}$$

applies to the whole operating range of the three-phase bridge

$$0 \leqslant \alpha \leqslant 180^{\circ}$$

i.e. for both the rectifier and the inverter. Its validity does not, however, extend to the condition of double commutation. The load characteristic area is shown in Figure 63. The limit of inverter control is determined by the condition

$$\alpha \leqslant 180^{\circ} - \mu \tag{58}$$

$$\cos(\mu + \alpha) = \cos\alpha - 2\,\mathrm{d}x \tag{59}$$

when $\mathrm{d}x = \dfrac{1}{\sqrt{3}}\, i_{\mathrm{d}} = \dfrac{1}{\sqrt{3}}\dfrac{I_{\mathrm{d}}}{I_{\mathrm{dk}}}$

From equations (57) to (59), the direct voltage of the inverter at the limit of control is

$$v_{\mathrm{dmax}} = -\,(1 - \mathrm{d}x) \tag{60}$$

which in absolute terms is equal to the rectifier voltage with $\alpha = 0$ (equation 43):

$$v_{\mathrm{dmax}} = 1 - \mathrm{d}x \tag{61}$$

The maximum permissible delay angle of the inverter (the limit of control) can be determined from equations (58) and (59):

$$\alpha_{\max} = \cos^{-1}\left(\frac{2}{\sqrt{3}}\, i_{\mathrm{d}} - 1\right) \tag{62}$$

These equations (60 to 62) are valid only for $\mu < 60^{\circ}$ (which is usually sufficient for the normal operating conditions of the converter).

COMPARISON OF RECTIFIERS AND INVERTERS

The controlled three-phase bridge converter can function either as a rectifier or as an inverter. The transition from one operating mode to the other can take place within one cycle (Figure 61) through shifting of the control pulses. This assumes that the back-voltage presents a corresponding polarity. The gate-pulse pattern (Figures 48 and 49) is unaltered.

The converter can operate as a rectifier with uncontrolled valves—diodes (Figures 27 to 47), but inverter operation can be achieved only with controllable valves—thyristors (Figures 48 to 63).

In the rectifier mode, active power is transferred to the d.c. circuit from the three-phase supply system. If the d.c. circuit includes a d.c. machine, it operates as a motor.

In the inverter mode the active power is transferred from the d.c. side to the three-phase supply system. In a d.c. drive the machine functions as a generator (the motor is braked).

The d.c.-side back-voltage can alternatively be provided from a second converter. This double-bridge circuit constitutes an intermediate-link frequency changer (see below). Here the converters operate alternately in the rectifier or the inverter mode according to the required direction of energy flow (one as a rectifier, the other as an inverter).

Theoretically the same direct voltage (in absolute terms) can be obtained

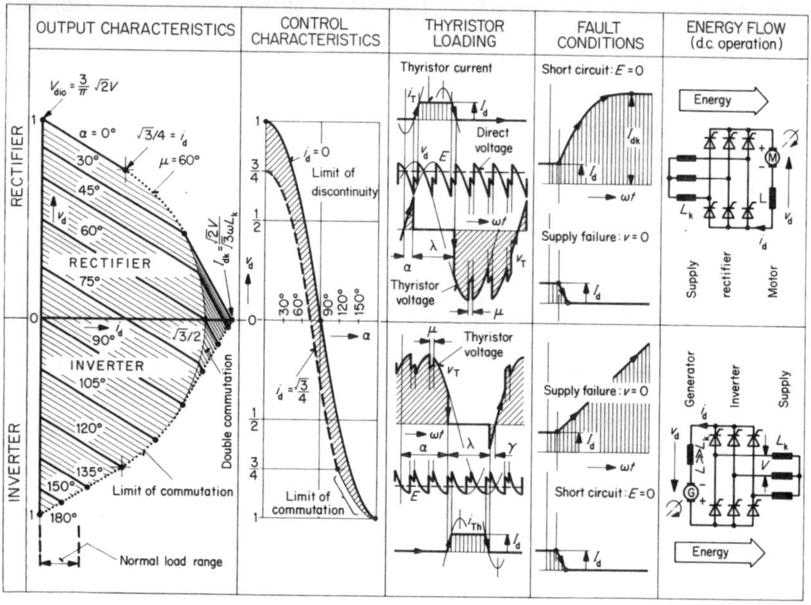

Figure 64. Comparison of rectifier and inverter operation in the three-phase bridge. Load and control characteristics. Typical voltage and current waveforms and typical fault situations. Operation illustrated by the example of a d.c. drive

from the converter in the inverter as in the rectifier mode. In practice, because of the necessary safety margin inside the limit of inverter control, the voltage obtainable from the inverter is lower than that of the rectifier. In both operating modes the thyristors are subjected to the same mean current and the same maximum voltage.

In the inverter mode the thyristor supports predominantly positive voltage, in the rectifier mode predominantly negative (see Figure 121).

The fundamental difference between inverter and rectifier operation in the converter arises under fault conditions (for details, see below).

In rectifier operation a d.c.–side short circuit results in an overcurrent, whose maximum value corresponds to the short-circuit current of the transformer ($I_{dk} = I_m$). The converter can, however, be blocked by triggering delay, and the short-circuit current suppressed.

In a rectifier a supply-voltage dip does not cause an overcurrent. In the case of an inverter, by contrast, a supply-voltage dip (even of short duration) is dangerous. The inverter loses control and the direct current driven by the d.c. source rises continuously and can no longer be influenced by gate control (commutation failure) (see Figures 108 to 110).

A d.c.-side short circuit on the inverter produces no overcurrent in the inverter.

A summary of the principal features of the rectifier and the inverter is presented in Figure 64.

CONVERTER OPERATION WITH DISCONTINUOUS DIRECT CURRENT

The direct current is continuous only if the current can flow for more than 60° after the initial turning-on of the valves. The conduction period of 60°, the *discontinuity limit,* is established when the back-voltage E is exactly equal to the mean value of the inverter terminal voltage V_d. The transition from discontinuous to continuous direct current in the case of the diode bridge is shown in Figure 34.

The analogous relationships in the controlled bridge are shown in Figure 65. So long as the back-voltage E is greater than the maximum converter voltage V_{dm} no current can flow. If the back-voltage E falls below the instantaneous converter voltage v_d, a current i_d is driven through the load circuit by the voltage–time integral A resulting from the difference between the two voltages. The current remains discontinuous so long as the conduction period $\lambda \leqslant 60°$. For any particular delay angle α a back-voltage E can be determined that will result in a discontinuous direct current (a small delay angle is assumed; see Figure 69).

The discontinuous current is dependent upon the load inductance L; the greater the circuit inductance $(2L_k + L)$, the smaller is the current. The individual current pulses are independent of one another, and the converter actually operates as six single-pulse converters (see Figure 15) in sequence. *When commutation is in evidence, however, the direct current i_d (under steady-state conditions) is independent of the load inductance.*

With very high load inductance, $L/L_k \to \infty$, the discontinuous current is

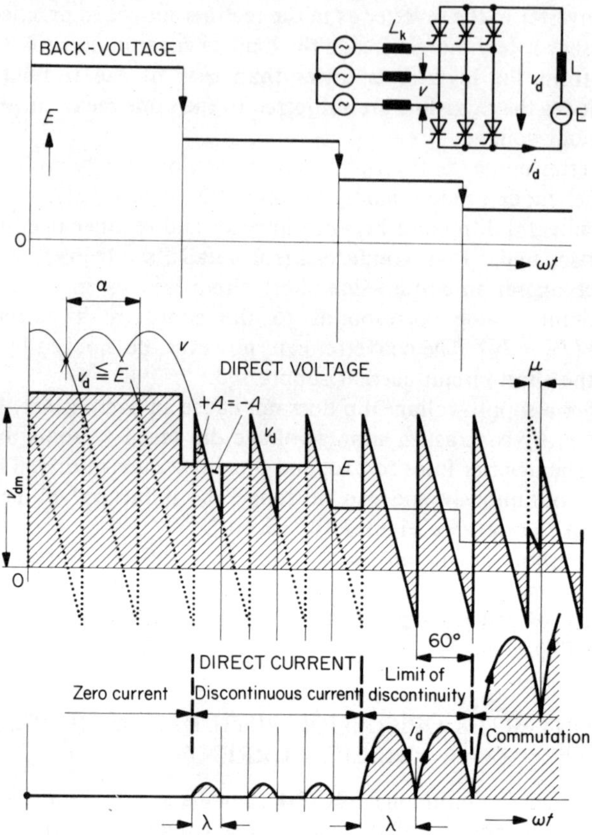

Figure 65. Transition from the zero-current condition in the three-phase bridge to the condition of continuous direct current. Diagrammatic representation for $\alpha = 75^\circ$ and step-wise variation of back-voltage. The limit of discontinuity is reached when $E = V_d = V_{dio} \cos \alpha$

very small and can usually be neglected (conventional representation). Discontinuous current must nevertheless often be taken into account, especially when the inductance ratio L/L_K is not very great. In the extreme case, $L = 0$, the discontinuity limit lies in the region of the rated load.

Figure 66 shows a section of the load diagram of the three-phase bridge, illustrating as an example the characteristics for $\alpha = 60^\circ$ and $\alpha = 120^\circ$ in the extreme cases with $L/L_K \rightarrow \infty$ and $L/L_K \rightarrow 0$. Also indicated in the diagram are the discontinuity limits for $L/L_K = 0$ and $L/L_K = 2$. It will be seen that at low currents the characteristics deviate from one another considerably. The limit discontinuous current is greatest when $\alpha = 90^\circ$ with a back-voltage $E = 0$. Figure 67 shows the relationships for $E = 0$ (as for a stationary machine in a d.c. drive). At the limit of discontinuity the direct voltage (mean) is zero, but a current nevertheless flows in the load. Only when $\alpha = 120^\circ$, when the converter voltage contains no positive voltage–time integral, does the current fall

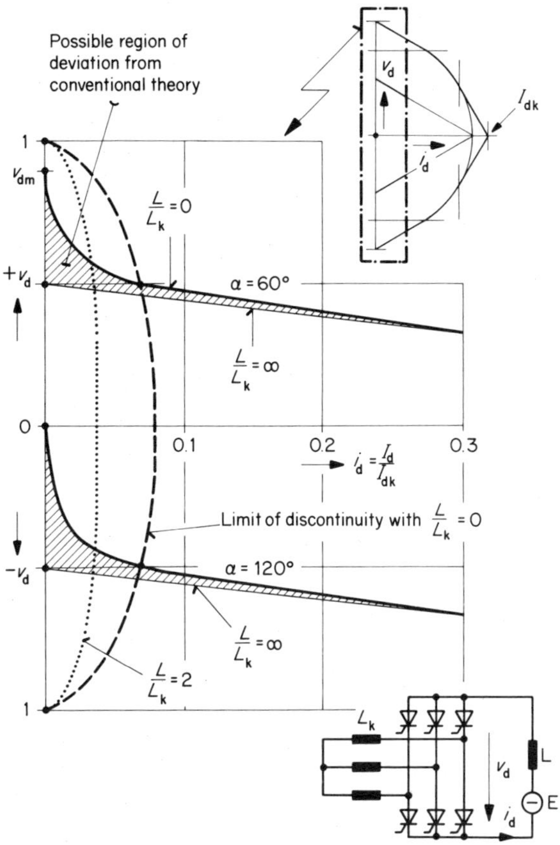

Figure 66. Load characteristics of the three-phase bridge in the region of discontinuous direct current. Example with $\alpha = 60°$ and $\alpha = 120°$ and no load inductance ($L = 0$). Maximum value of mean direct current at the limit of discontinuity (extreme case with $L/L_K = 0$). Characteristics for $L/L_K = \infty$ for comparison

to zero. This demonstrates that *with a purely inductive load circuit (with no back-voltage) the converter is completely blocked—i.e. non-conducting—only when* $\alpha > 120°$.

With a very small delay angle ($\alpha < 10°$) discontinuous current is not possible. This is shown in Figure 68. In this region of control the thyristors can only be triggered initially at the intersection of the two voltages (v, E). The converter then goes straight into the continuous current condition.

The discontinuity limit with a delay angle $\alpha = 0°$ is represented by $E/V_{dio} = 1$. However, since the instantaneous voltage v is less than E, the converter can not in that case be triggered. If, on the other hand, the initial triggering is effected by, for example, a wide gate pulse (12.7°), the operating point can then be driven further towards the discontinuity limit for $\alpha = 0°$. The behaviour of the converter in the region of small delay angles and high back-voltage cannot therefore be positively defined.

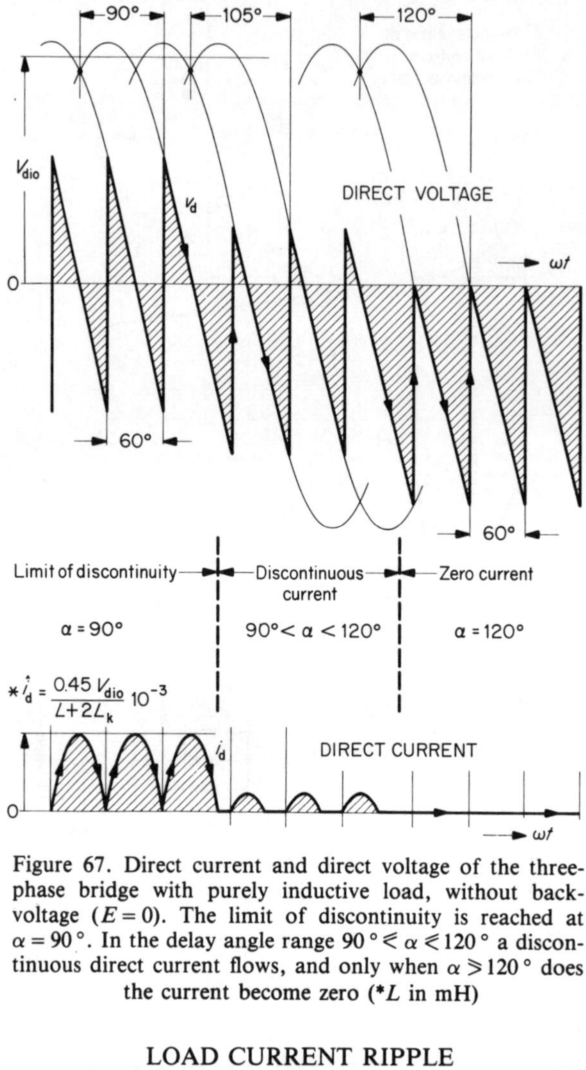

Figure 67. Direct current and direct voltage of the three-phase bridge with purely inductive load, without back-voltage ($E = 0$). The limit of discontinuity is reached at $\alpha = 90°$. In the delay angle range $90° \leqslant \alpha \leqslant 120°$ a discontinuous direct current flows, and only when $\alpha \geqslant 120°$ does the current become zero (*L in mH)

LOAD CURRENT RIPPLE

The largest fluctuation of the load current in the three-phase bridge occurs when $\alpha = 90°$ and $E = 0$. The direct current in this case is at the limit of discontinuity, and is dependent upon the circuit inductance $L + 2L_k$ (see Figures 66 and 67). The smaller the rectifier delay angle, for a given ratio L/L_k, the smaller is the fluctuation of the direct current i_d. The ripple decreases with increasing L/L_k (see Figure 37), and with a very large L/L_k the direct current is completely smooth. The direct current fluctuations are determined by the voltage–time integral resulting from the difference between the converter voltage v_d and the back-voltage E. The difference v_L between the two voltages appears across the load inductance L. The inductor voltage v_L is a purely alternating voltage (with a mean value of zero) whose fundamental component is of six times the supply frequency. An alternating voltage is developed across

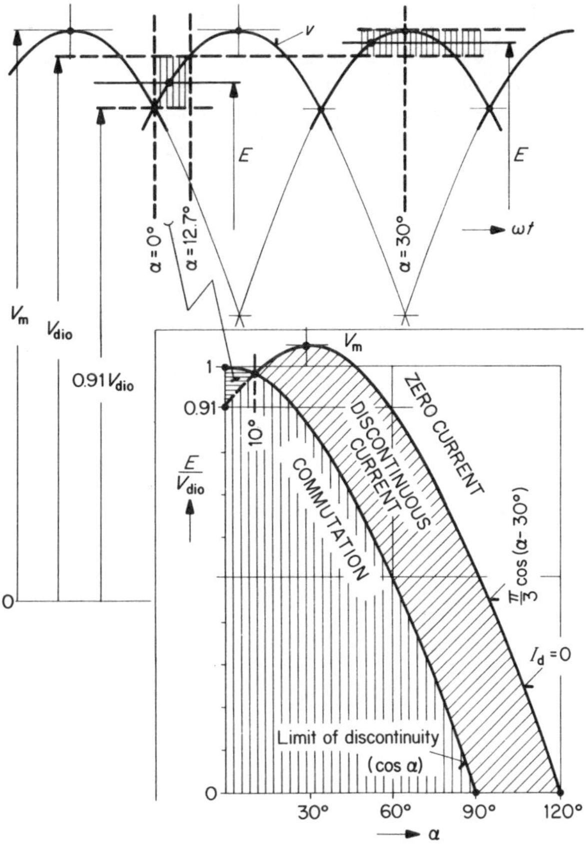

Figure 68. With small delay angles, $0° \leqslant \alpha \leqslant 10°$, the converter invariably goes straight into the continuous current operating mode after triggering. In the range of voltages $0.91 \leqslant E/V_{dio} \leqslant 1$ either there is no current or the converter commutates. Control characteristics and example

the load inductance even though the current is assumed to be completely smooth. The inductor voltage is independent of the ripple current. The ripple current is larger in the d.c.-link converter (See Figures 133 to 135).

MATHEMATICAL APPENDIX XII

The following current and voltage relationships are derived in accordance with the inductance law:

$$v_L = L \frac{di}{dt} \tag{63}$$

$$i = \frac{1}{L} \int v_L \, dt = \frac{1}{L} A \tag{64}$$

where $A = \int v_L \, dt$ (voltage–time integral)

With a positive voltage–time integral the current increases; with a negative it falls. The positive voltage–time integral A across the load inductance of the three-phase bridge can be calculated, for a given delay angle α and overlap angle μ (or for corresponding direct voltage and current), as the difference between the direct voltage and the back-voltage:

$$v_L = v_d - E$$

so that by means of equation (64) the current fluctuation can be determined as follows:

$$\Delta i_d = \frac{1}{L} A \tag{65}$$

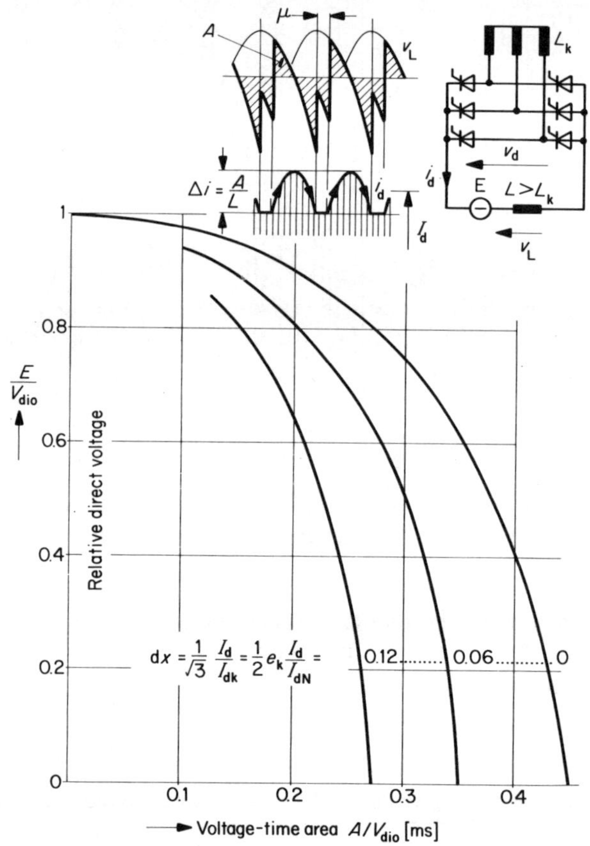

Figure 69. Diagram for the determination of the fluctuation Δi of the load current in the three-phase bridge. A = voltage–time integral, e_k = transformer short-circuit voltage, I_{dN} = rated converter current, dx = relative inductive d.c. voltage drop, I_{dk} = short-circuit direct current (approximation)

Figure 69 gives the voltage–time integral, relative to V_{dio}, with dx as parameter, as a function of the relative direct voltage $v_d = E/V_{dio}$. This diagram can be used to estimate the ripple current, albeit only approximately.

THE PRINCIPAL CONVERTER THEORIES

According to the purpose, three different theories may be applied to circuit analysis in converter engineering. The main difference between these theoretical representations lies in the complexity of the equivalent circuits.

The *idealized theory* (see Figures 29 to 32 and 49 to 51) deals with an ideal, non-inductive three-phase system and smooth direct current. Commutation takes place infinitely quickly, the valve currents are of rectilinear waveform, and the characteristics embody no voltage drop. This simplification is adopted for rough calculations and, mainly, for estimating current ratings of the valves and transformers.

The *conventional theory* (see Figures 39, 41 to 44, and 54 to 58) also assumes a completely smooth direct current, but extends to a three-phase system with commutating inductances. As a result of commutation the valve currents are of trapezoidal waveform. The load characteristics are drooping and double commutation is introduced as the short-circuit current is approached. This theory is mostly used in analytical studies of the three-phase bridge, but only in the steady state. It is most usefully employed in connection with the voltage

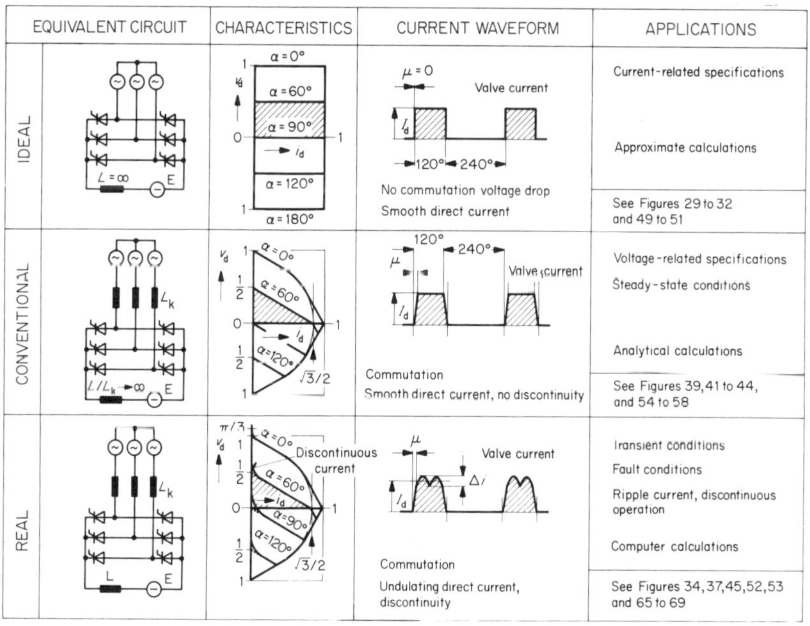

Figure 70. In converter studies, three different theories are applied according to the requirements. The difference lies in the various assumptions as to the ratio of circuit inductances: $L/L_k = 0$ in the idealized theory, $L/L_k = \infty$ in the conventional theory, and $0 < L/L_k < \infty$ in the real theory

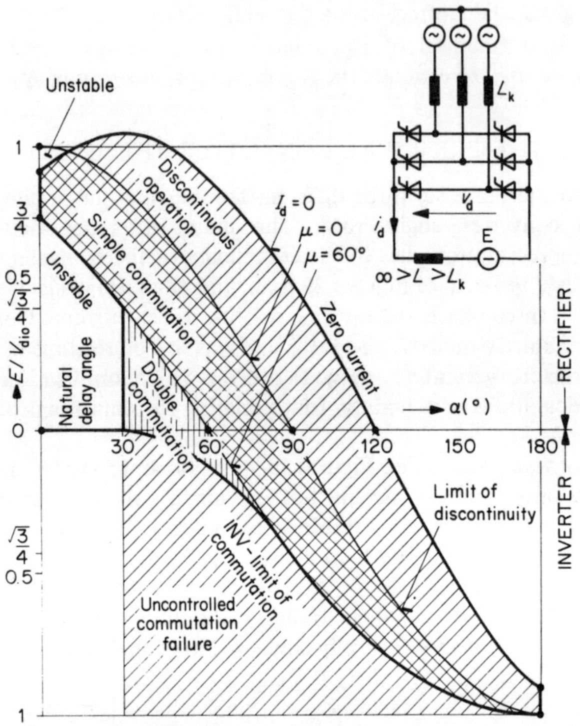

Figure 71. Possible operating conditions of the controlled three-phase bridge with real L/L_k ratios. The delay angle α and the relative load back-voltage E/V_{dio} determine the corresponding steady-state condition of the converter (inverter–rectifier, zero-current–discontinuous-current–continuous current –unstable–loss of control)

ratings of the converter and the converter transformer, and for the calculation of supply loading.

In the conventional theory the d.c. inductance is assumed to be infinitely large. This permits simple steady-state calculations, but precludes the study of transient processes (the very high inductance prevents any variation of the current). If a finite load inductance is introduced, as in the *real theory,* the steady-state conditions are more difficult to calculate, but the transient processes, fault conditions, dynamic performance, and the discontinuous current region become amenable to calculation.

To facilitate comparison, these three theories of the three-phase bridge are displayed in Figure 70. In Figure 71 all the possible operating conditions of the converter with a finite inductance ratio, $\infty > L > L_k$, are illustrated by means of the control characteristics. It will be seen that the converter can operate, in both the rectifier and the inverter modes, in the discontinuous current region, with simple commutation or with double commutation. Which (steady-state) operating mode applies is decided only by the delay angle and the back-voltage. The current is then determined according to the load characteristics

(Figures 56, 59, and 63). The unstable region depends upon the nature of the triggering pulses. With diodes there is a natural variation of the delay angle (Figures 41 and 42), whereas with thyristors with narrow triggering pulses there is a jump to the characteristic for $\alpha = 60^\circ$ (Figures 52 and 53).

8

Power Balance in Three-Phase Bridge Converters

GENERAL

The converter functions as a *power regulator*. In circuit theory the converter is considered as a lossless regulator, whose function is to translate a desired active power (and, in special cases, reactive power) to the load. The translation is indeed lossless, but, because of control and commutation, non-active power is introduced. In the operation of the three-phase bridge the three-phase system (the supply, transformer, or machine) has to be capable of supplying an apparent power correspondingly greater than the active power.

The parameters applicable to the power balance in converter engineering are the *power factor,* the *displacement factor,* and the *distortion factor.* The basic concepts of power balance are illustrated in Figures 20 to 22. Contrary to the practice with sinusoidal currents it is necessary in the case of converters to take into account the distorted current waveform and also the *dynamic conditions.* Special concepts are applied here which define the mathematical–physical procedures. Thus as an example the concept of non-active power is extended, and a distinction is drawn between *fundamental frequency reactive power, control reactive power, commutation non-active power,* and *distortion non-active power.*

CONTROL REACTIVE POWER OF A CONVERTER

In an ideal three-phase bridge converter the direct current is completely smooth and the alternating current is of rectilinear waveform (Figures 50 and 70). Power is translated between the a.c. voltage source and the d.c. voltage source, and the power on the d.c. side is equal to the active power in the three-phase system. The active power in the a.c. voltage source results only from the *fundamental components* of the voltage and current and their *displacement factor* (Figures 20 and 22).

The r.m.s. value of the fundamental component is always less than (or, at the most, equal to) the total r.m.s. value of the waveform in question. In an ideal three-phase bridge the fundamental component I_1 of the alternating current is less than the r.m.s. value I by a factor of $3/\pi$ (i.e. by about 5 per cent.).

The phase displacement ϕ_1 between the fundamental component I_1 of the current and the alternating voltage V (an ideal sinusoidal voltage is assumed, so that $V = V_1$) in an ideal converter is equal to the delay angle α. Figure 72 shows, for $\alpha = 30^\circ$ as an arbitrary example, the corresponding direct voltage

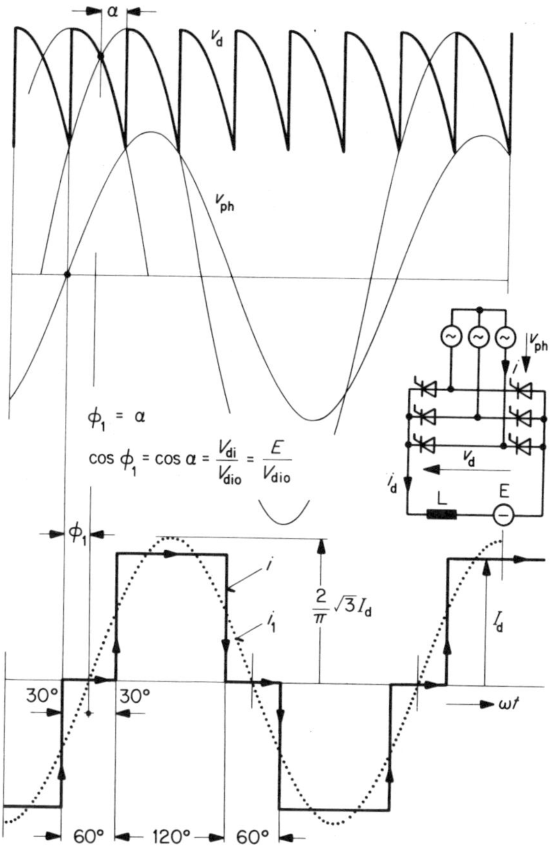

Figure 72. In the ideal three-phase bridge the phase shift ϕ_1 of the fundamental alternating current i is equal to the delay angle α. Example with $\alpha = 30°$. Top: direct voltage v_d. Bottom: alternating current i and its fundamental component i_1. The active power is given by $P = EI_d = V_{dio}I_d \cos\alpha = V_{ph}I_1 \cos\phi_1 \times 3$

v_d and the alternating current i in an ideal converter. Relative to the phase voltage v_{ph}, the fundamental component of the current is phase-shifted by an angle $\phi_1 = \alpha$. Control of the delay angle varies the direct voltage V_d and hence the active power ($P = V_dI_d$). The alternating current waveform is not changed by the control, and its r.m.s. value I (with a constant direct current) therefore remains constant. The loading on the three-phase system—the apparent power S—is thus not affected by the delay angle α (assuming constant direct current). When $\alpha = 90°$ the active power is zero, but the supply is still fully loaded, and the whole of the translated power is in the form of non-active power Q. The ratio of active to apparent power, P/S, is termed the *power factor* λ. The greater the power factor λ, the better is the three-phase system utilized.

The *total apparent power* in the three-phase system is given by the r.m.s. value of the current ($S = \sqrt{3}\ VI$). The related *total non-active power*

($Q = \sqrt{S^2 - P^2}$) is also calculated from the r.m.s. current I. Calculation based upon the r.m.s. fundamental current I_1 gives the *fundamental frequency apparent power* S_1 and the *fundamental frequency reactive power* Q_1. The ratio P/S_1 is the *displacement factor* $\cos \phi_1$.

Since in an ideal converter $\cos \phi_1 = \cos \alpha$, the fundamental frequency reactive power, $Q_1 = S_1 \sin \phi_1$, may be designated the *control reactive power*. With constant direct current I_d the active power P and the reactive power Q_1 vary with control, but the fundamental frequency apparent power S_1 remains constant. In a graphical representation the apparent power S_1 thus describes a semicircle (Figure 73). The active power P is positive in the rectifier mode and negative in the inverter mode. The reactive power Q_1 is always lagging. At the point $P = 0$ the semi-circle is interrupted, since this condition cannot physically exist.

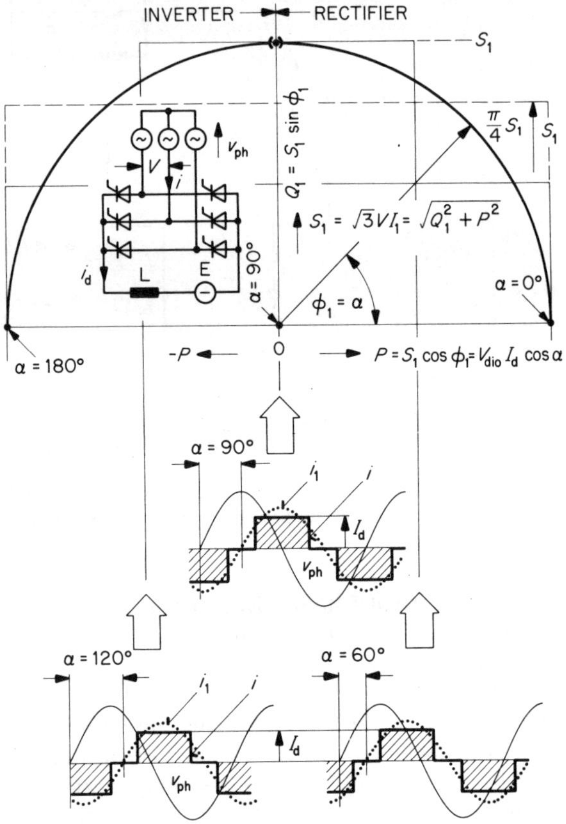

Figure 73. In the ideal converter with constant direct current I_d the fundamental frequency apparent power $S_1 = \sqrt{3}\ VI_1$ describes a semicircle with varying delay angle. The delay angle α is equal to the phase angle θ_1. The fundamental frequency reactive power Q_1 is termed the control reactive power

COMMUTATION NON-ACTIVE POWER

As a result of *commutation* (Figures 35 and 60) the valve and alternating currents acquire a trapezoidal waveform. The distortion of the current waveforms depends upon the delay angle α and the load. The commutation angle is largest when $\alpha = 0^\circ$ and smallest when $\alpha = 90^\circ$. With increasing direct current I_d the alternating current departs progressively further from the ideal rectilinear waveform (to the point of becoming a pure sine wave on short circuit; see Figure 45). Figure 74 shows the alternating current i with $\alpha = 0^\circ$ for two arbitrary overlap angles, $\mu = 30^\circ$ and 60°. Also given in Figure 74 are the fundamental component I_1/I and the ratio of the r.m.s. current I to the r.m.s. value of the ideal rectilinear wave as functions of overlap angle μ.

Commutation modifies not only r.m.s. fundamental component I_1 but also its phase angle ϕ_1. This is shown in Figure 75. In comparison with the

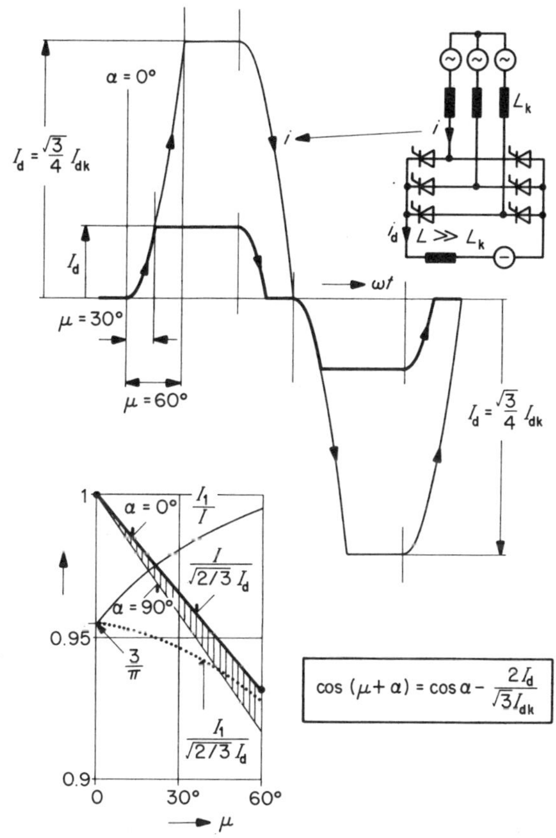

Figure 74. Commutation alters the waveform of the alternating current i. With increasing overlap angle μ the fundamental component I_1 increases as a proportion of the total r.m.s. current I. The r.m.s. value I relative to the ideal rectilinear wave current is reduced. Examples with $\alpha = 0^\circ$ $\sqrt{2/3}\ I_d$ = r.m.s. value of rectilinear wave current

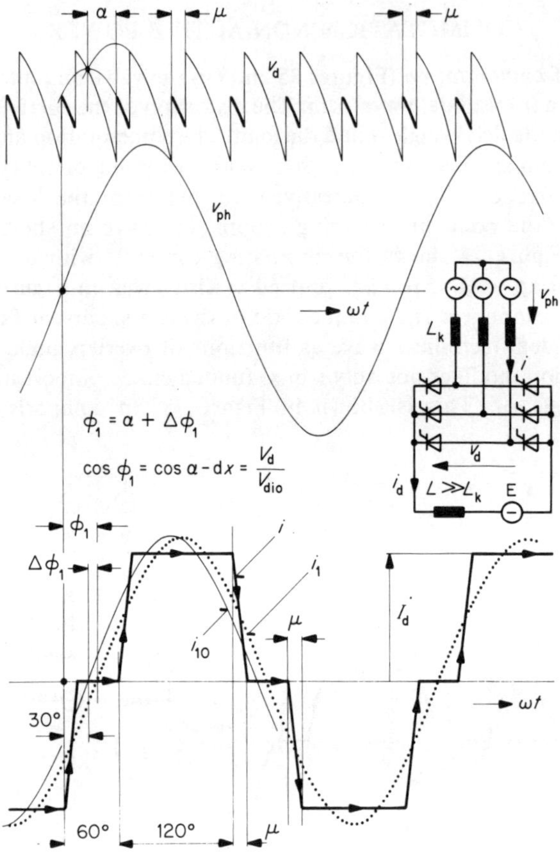

Figure 75. Due to commutation the fundamental component i_1 of the alternating current i is phase-shifted by an angle $\Delta\phi_1$. The phase shift ϕ_1 is greater than in the ideal case of a rectilinear wave current (Figure 72) because of the overlap angle μ

rectilinear waveform (Figure 72) the phase angle is increased by $\Delta\phi_1$ due to commutation. The phase shift ϕ_1 is now greater than the delay angle α.

The fundamental component of current is phase-displaced relative to the voltage (ϕ_{10}), even when $\alpha = 0^\circ$. The converter therefore loads the three-phase system with reactive power even at maximum output. Since this reactive power originates from commutation, it can be described as *commutation reactive power*.

A representation of the division of the fundamental frequency reactive power into control and commutation components is given in Figure 76. The significance of commutation reactive power in the whole control range of the rectifier can be seen from an arbitrary example in the load diagram ($dx = 0.05$). The diagram—the relationship of reactive to active power—is scaled in relative values, with the fundamental frequency apparent power as the reference quantity. The relative active power p is equal to the relative direct voltage

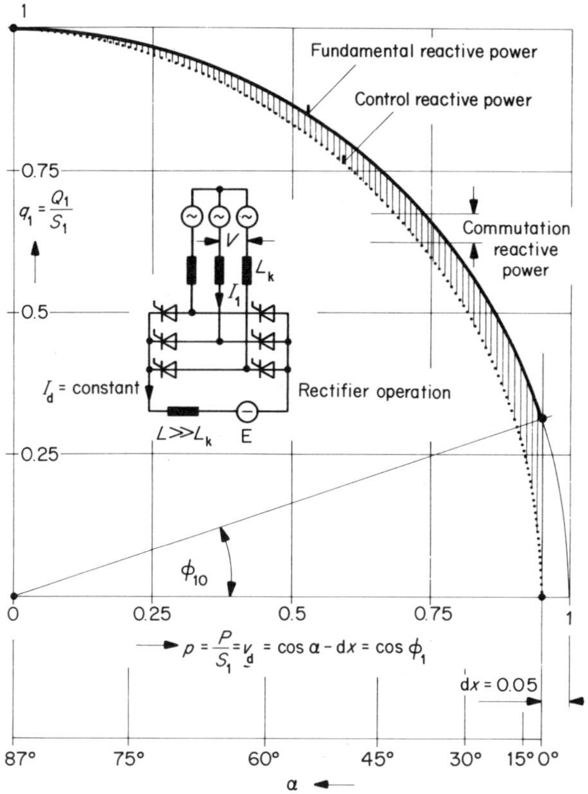

Figure 76. The fundamental frequency reactive power Q_1 in the three-phase bridge is a combination of control and commutation reactive power. Example of the division with $dx = 0.05$ in the rectifier mode. Relative values (reference value: fundamental frequency apparent power $S_1 = \sqrt{3}\, VI_1$).). ϕ_{10} = phase shift with $\alpha = 0°$

$v_d = V_d/V_{dio}$, or alternatively to the displacement factor $\cos\phi_1$. The diagram is valid only for steady-state conditions, for smooth and constant direct current I_d and in the region of simple commutation.

The correspondence between the power circle diagram and the load diagram of the three-phase bridge is illustrated in Figure 77 (without subdivision of the fundamental frequency reactive power).

DISTORTION POWER

The phase shift ϕ_1 of the fundamental frequency current I_1 is due to control and commutation. As a result, the converter loads the supply system with reactive power (the fundamental component). Even with an ideal converter, however, without commutation and giving full output (Figures 30 and 50, for example), the a.c. voltage source has to supply more than the active power.

The active power in the a.c. voltage source is due only to the fundamental component of the current (see Figure 22). If the current is not sinusoidal,

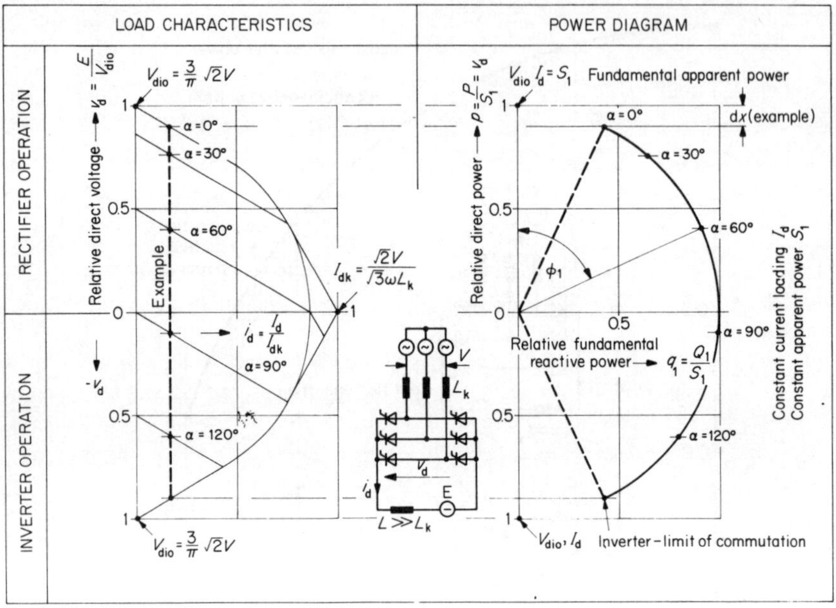

Figure 77. The power circle diagram (right) represents the power relationships in the converter, for a given constant loading, over the whole range of load diagrams (left). Example with $dx = 0.1$. Steady-state conditions, large inductance ratio L/L_k. Discontinuous current region ignored

which is generally the case in a converter, the non-sinusoidal component of the current (the harmonics) also loads the supply system, but without contributing to the active power transmission. This loading can be considered as a kind of non-active power. It can be termed the ***harmonic non-active power, or the distortion power.***

With increasing load current I_d the distortion power Q_H decreases, relative to the apparent power S_1, since the fundamental component I_1/I of the alternating current increases (Figure 74). The effect of the delay angle α on the distortion power (with constant direct current) is insignificant.

Because of the distortion power Q_H the total non-active power Q in the three-phase bridge is at the most 5 per cent. greater than the fundamental frequency reactive power Q_1. The non-sinusoidal supply current therefore makes the power factor $\lambda = P/S$ worse, but has little direct influence on the displacement factor $\cos \phi_1$.

GEOMETRICAL REPRESENTATION OF THE CONVERTER POWER BALANCE

The following power quantities are used in converter theory:

Active power:	P
Apparent power:	total apparent power S
	fundamental frequency apparent power S_1

Non-active power:	total non-active power Q
	fundamental frequency reactive power Q_1
	control reactive power
	commutation reactive power
	distortion power Q_H
Power factor:	$\lambda = P/S$
Displacement factor:	$\cos \phi_1 = P/S_1$

As in conventional a.c. theory, here also the power relationships can be represented in a *phasor diagram* (Figure 21). However, only the fundamental frequency powers (S_1, Q_1, P) can be represented by phasors. The total apparent and non-active powers, S and Q, and the distortion power, Q_H, are only numerical values.

The distortion power Q_H is perpendicular to the plane P, Q_1, S_1, so that the total apparent power S is represented geometrically by a *space diagonal.* Figure 78 shows, above, the powers in one plane, and at the bottom in three-dimensional form.

All the above considerations relate to the power relationships at the a.c.

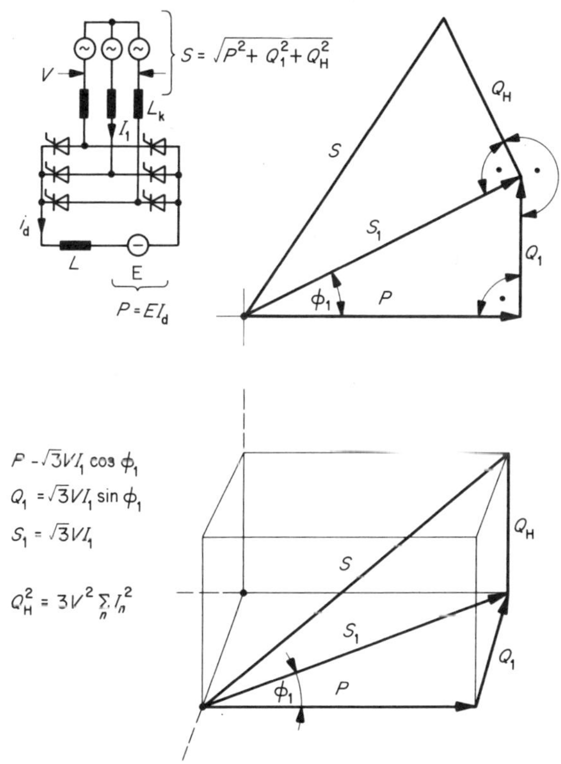

Figure 78. The power relationships in an a.c. voltage source supplying a converter can be represented geometrically in space (bottom) or opened out into a plane (top). Q_H = distortion power, I_n = nth harmonic current ($n = 5, 7, \ldots$)

voltage source. *Directly on the a.c. terminals* of the converter the power balance has a different appearance. The current I is the same here as in the voltage source v, but the voltage V_{sec} is no longer an ideal sine wave, being distorted by the voltage drops v_{Lk} across the commutating inductances L_K.

The fundamental component V_{S1} of the alternating voltage on the converter is therefore less than the source voltage V and embodies a phase shift ϕ_S.

Figure 79 shows graphically the current and voltage waveforms on the a.c. side of the converter. The corresponding phasor diagram is valid only for the fundamental components. The fundamental component v_{Lk1} of the voltage drop is phase-shifted 90° relative to the fundamental current I_1 and, together with the fundamental component V_{S1} of the phase voltage, makes up the source voltage V_{ph}. (See also the phasor representation of Figure 136.)

Figure 80 shows the variation of the fundamental currents and their phase angles with the direct current loading. It should be emphasized that the active

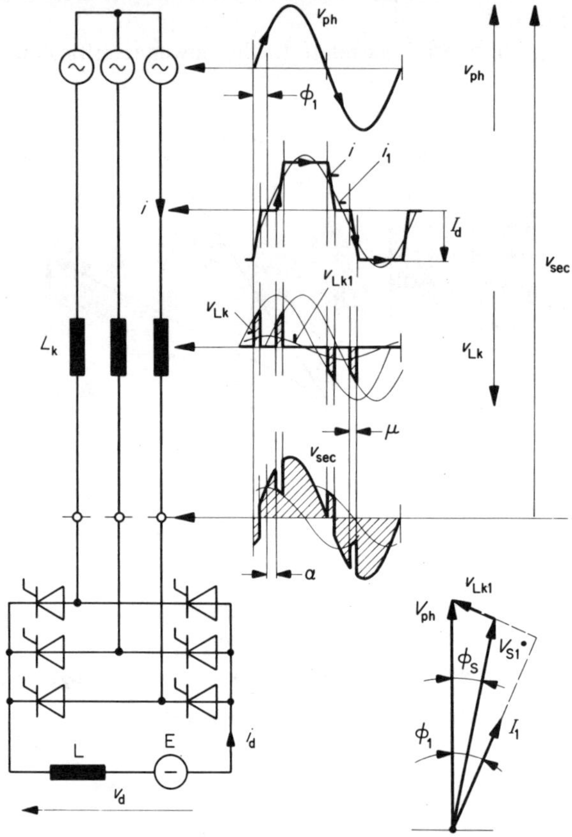

Figure 79. The voltage v_s at the a.c. terminals of the converter is distorted as a result of commutation, and the fundamental component v_{S1} lags the source voltage by an angle ϕ_S. The fundamental component of the inductive voltage drop v_{Lk1} leads the fundamental current I_1 by 90 °

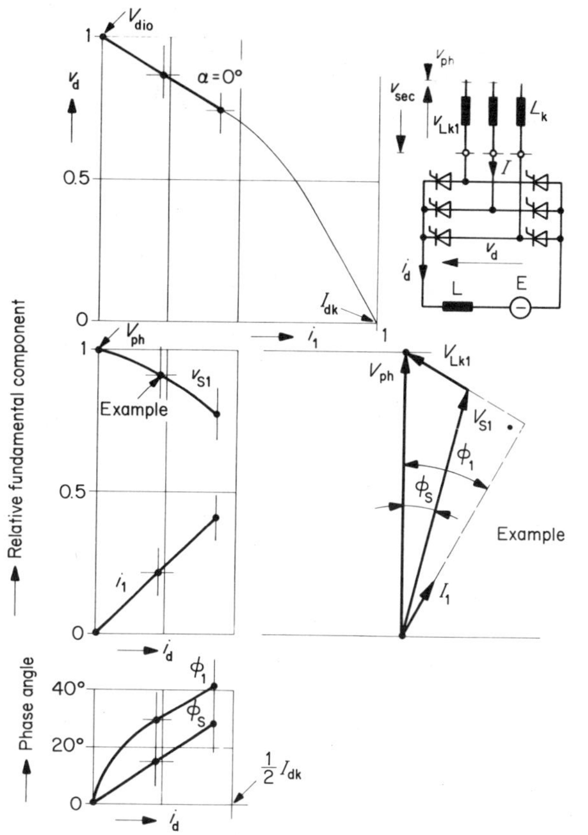

Figure 80. With increasing current loading i_d of the converter the fundamental component v_{S1} of the phase voltage falls and is phase-shifted by ϕ_S relative to the source V_{ph}. The fundamental alternating current I_1 is also phase-shifted by commutation (ϕ_1). Relationships with $\alpha = 0°$. Top: load diagram. Right: phasor representation.

power at the a.c. terminals of the converter is the same as that in the a.c. voltage source and in the d.c. voltage source.

Figure 81 shows how the active power and the fundamental frequency reactive power vary in dependence upon the load current over the whole range of rectifier operation and for three arbitrary delay angles. It will be seen that the active power goes through a maximum, albeit outside the normal load range (reference quantity: $V_{dio}\ I_{dk}$).

POWER RELATIONSHIPS IN THE TRANSIENT STATE

Theoretical studies of the three-phase bridge are undertaken with the aid of an equivalent circuit (Figure 70), which actually relates to a d.c. converter drive. Electric motor drives, however, frequently operate under transient conditions.

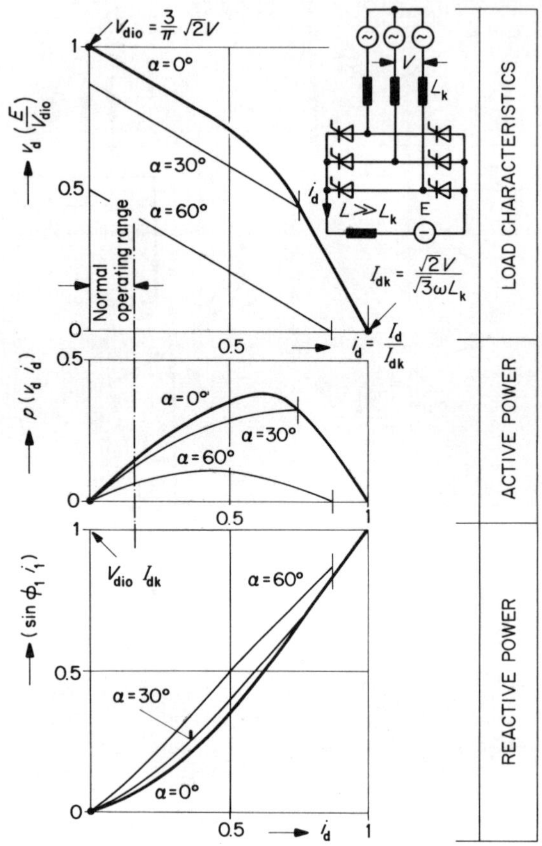

Figure 81. With increasing current loading i_d the fundamental frequency reactive power in the three-phase system increases continuously. The active power, on the other hand, goes through a maximum. Three-phase bridge in the rectifier mode with delay angle α as parameter. Steady-state conditions. Reference values for p and q_1: $V_{dio} \times I_{dk}$

During acceleration, braking, and variations of speed or load, the direct voltage and current of the converter vary accordingly.

The following three typical transient states of the converter can be distinguished:

1. Variation of current with constant delay angle
2. Variation of the direct voltage with constant current loading
3. Simultaneous variation of direct current and voltage

For transient conditions that do not change very quickly (within one cycle), the same approach to power balances can usually be adopted as for the steady state (Figures 72 to 81). In the case of rapid changes, however, the following difficulties arise:

The frequency f_1 of the fundamental current does not coincide with the frequency f_v of the voltage source.
The individual phases of the three-phase system are not symmetrically loaded.

As an example, Figure 82 shows the currents i_R, i_S, and i_T in the individual phases of the three-phase system at the beginning of a braking period in a d.c. drive. It will be seen that the current loading of the phases in the first cycles of the process is quite unbalanced and that the current waveform is very distorted. To be able to calculate the powers in the first cycle, it is necessary to ascertain the total r.m.s. values and fundamental components of the individual currents during this time interval. Then the apparent, active, and non-active powers of the individual a.c. voltage sources can be determined. (In

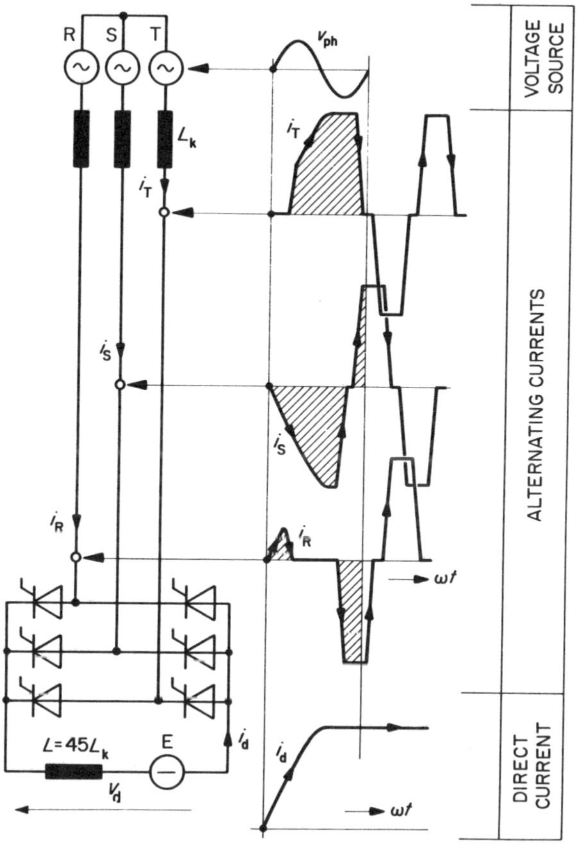

Figure 82. In transient conditions of the converter the current loadings of the individual phases of the three-phase system can be very different. The diagram shows the transition into braking operation in a d.c. drive. The fundamental frequency reactive power Q_1 in the first cycle is different in the individual phases R, S, T. (Example)

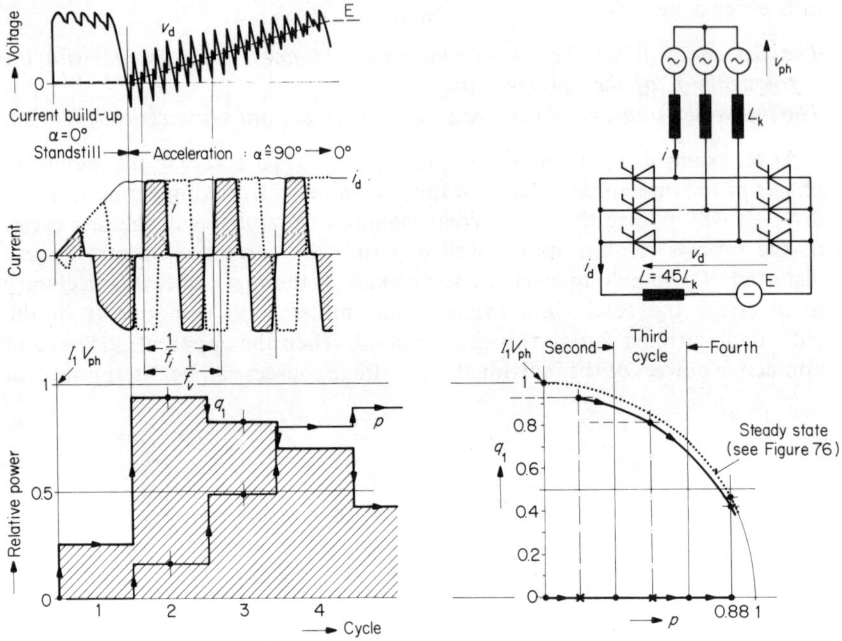

Figure 83. During acceleration in a converter drive the frequency f_1 of the alternating current is less than the supply frequency f_v. The fundamental frequency reactive power q_1 calculated in the individual periods lies under the values given by the steady-state power diagram (Figure 76). Example

the first example in Figure 82 the fundamental frequency reactive power in phase R for the first cycle amounts to only 60 per cent. of the reactive power in phase T.)

During acceleration in a d.c. drive (see Figure 83) the delay angle is continuously shifted from $\alpha = 90°$ to $\alpha = 0°$. The direct current i_d is thereby held constant. Since the triggering pulses are continuously shifted and the conduction period consequently reduced, the frequency f_i of the alternating current i is higher than the frequency f_v of the voltage during the acceleration period. If the relative active power p and the relative fundamental frequency reactive power q_1 in the individual cycles are calculated for each of the three phases, the result is the graph shown in Figure 83 (bottom). Compared to the value given by the steady-state power diagram (Figure 76) the reactive power q_1 during acceleration is reduced. *During an m-cycle acceleration period the mean reactive power q_1 is less than that given by the steady-state diagram* (Figure 73). This is shown graphically in Figure 84 (with the simplification that α is assumed equal to ϕ_1). The longer the acceleration lasts, the smaller is the mean reactive power.

MATHEMATICAL APPENDIX XIII

The power on the d.c. side of the converter is

$$P = EI_d = V_d I_d \tag{66}$$

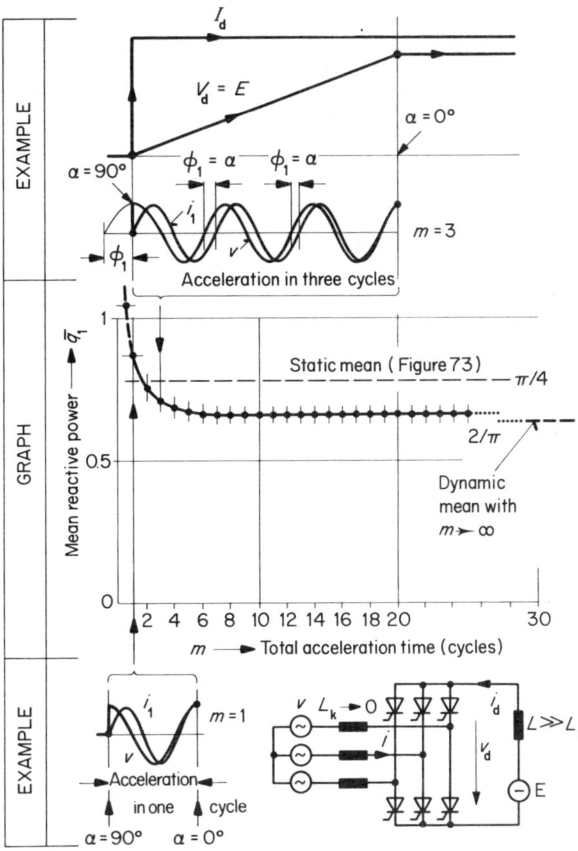

Figure 84. The mean reactive power $\bar{q}_1$ drawn by the converter during acceleration from $\alpha = 90°$ to $\alpha = 0°$ is less than that corresponding to the steady-state power diagram (Figure 73) (commutation reactive power neglected). m = total acceleration time in cycles. Example

The direct voltage is given by (equation (51))

$$V_d = V_{dio}(\cos \alpha - dx) \tag{67}$$

Therefore

$$P = V_{dio}\, I_d(\cos \alpha - dx) \tag{68}$$

Introducing the concept of *ideal d.c. power,*

$$P_{dio} = V_{dio}\, I_d \tag{69}$$

from which

$$P = P_{dio}(\cos \alpha - dx) \tag{70}$$

The three-phase system supplies the same amount of power. Hence

$$P = VI_1 \sqrt{3} \cos \phi_1 = S_1 \cos \phi_1 \tag{71}$$

where S_1 = fundamental frequency apparent power
I_1 = fundamental component of the alternating current

For the ideal rectilinear wave current of the three-phase bridge the fundamental component is

$$I_1 = \frac{1}{\pi} \sqrt{6}\, I \tag{72}$$

This falls slightly with overlap μ (Figure 74). From the equality of power (equations (70) and (71)),

$$P_{\mathrm{dio}} = S_1 \tag{73}$$

$$\cos \alpha - \mathrm{d}x = \cos \phi_1 \tag{74}$$

In the ideal converter without commutating inductance,

$$\cos \alpha = \cos \phi_1 \qquad \text{and } \alpha = \phi_1 \tag{75}$$

The fundamental frequency reactive power is

$$Q_1 = S_1 \sin \phi_1 \tag{76}$$

The powers form a right-angled triangle:

$$P^2 + Q_1^2 = S_1^2$$

From equations (71) and (76),

$$S_1^2 \cos^2 \phi_1 + S_1^2 \sin^2 \phi_1 = S_1^2$$

Therefore

$$\cos^2 \phi_1 + \sin^2 \phi_1 = 1 \tag{77}$$

Equation (77) is that of a circle, and the power balance of the converter can thus be represented by a circle diagram (Figures 73, 76, and 77).

The relative powers are as follows:

$$p = \frac{P}{S_1} = \frac{V_\mathrm{d}}{V_{\mathrm{dio}}} = v_\mathrm{d} = \cos \phi_1 \tag{78}$$

and

$$q_1 = \frac{Q_1}{S_1} = \sin \phi_1 \tag{79}$$

Here $\cos \phi_1$ is termed the 'displacement factor'. The total apparent power of the three-phase system is derived from the total r.m.s. alternating current I:

$$S = \sqrt{3}\, VI \tag{80}$$

For the rectilinear wave,

$$I = \sqrt{\frac{2}{3}}\, I_\mathrm{d}$$

or

$$I = \frac{\pi}{3} I_1 \tag{81}$$

The fundamental component I_1/I is increased by overlap (Figure 74). Fourier analysis of the rectilinear wave current of the three-phase bridge gives

$$I^2 = I_1^2 + I_5^2 + I_7^2 + I_{11}^2 + I_{13}^2 + \cdots \tag{82}$$

$$I^2 = I_1^2\left(1 + \frac{1}{5^2} + \frac{1}{7^2} + \frac{1}{11^2} + \frac{1}{13^2} + \cdots\right) \tag{83}$$

Here $I_5, I_7, \ldots, I_{13}, \ldots$ are the harmonics of the fifth, seventh, etc., orders. This can be generally written:

$$I = \sqrt{I_1^2 + \sum_n I_n^2} \qquad \text{for } n = 5, 7, 11, 13, \ldots \tag{84}$$

Multiplying by $\sqrt{3}\,V$ gives

$$\sqrt{3}\,IV = \sqrt{3V^2 I_1^2 + 3V^2 \sum_n I_n^2}$$

or, from equations (80) and (71),

$$S\sqrt{S_1^2 + Q_H^2} \tag{85}$$

where $Q_H = \sqrt{3}V\sqrt{\sum_n I_n^2}$ = distortion power (86)

Applying equation (77) gives

$$S = \sqrt{P^2 + Q_1^2 + Q_H^2} = \sqrt{P^2 + Q^2} \tag{87}$$

where $Q = \sqrt{Q_1^2 + Q_H^2}$ = total non-active power (88)

The total apparent power S can be considered, according to equations (85) and (87), as a *space diagonal* (Figure 78).

The ratio

$$\lambda = \frac{P}{S} \tag{89}$$

is termed the 'power factor'.

All the above equations are valid for steady-state conditions and for smooth direct current ($L \gg L_k$).

9

Harmonics in the Three-Phase Bridge Converter

GENERAL

Any non-sinusoidal periodic waveform can be considered as a combination of sine waves of certain frequencies, amplitudes, and phase angles. The mathematical proof, together with the relative theory, is supplied by *Fourier analysis.* According to Fourier analysis, any periodic time function can be resolved into a sine wave of the same frequency and an (infinite) series of sine waves of higher frequency, so that at any instant the sum of the instantaneous values of all the components gives the corresponding value of the analysed wave.

The sine wave of the same frequency is termed the *fundamental component.* The sine waves of higher frequency are known as *harmonics.* The harmonics are often designated according to their *orders n*, signifying the ratio of their frequencies to that of the fundamental component (so that, for example, in a 50 Hz system the 250 Hz component is referred to as the 'fifth' harmonic, since $250/50 = 5 = n$).

Fourier analysis is a mathematical process and the harmonics are therefore to be considered as an abstraction. However, it is relatively simple to demonstrate the physical significance of Fourier analysis experimentally. If, for example, a series of series-resonant L–C circuits, tuned to individual integral-multiple frequencies, are connected to a non-sinusoidal current source, it can be demonstrated that each resonant circuit carries only a sinusoidal current of the corresponding frequency and that it is of exactly the magnitude predicted by the theory. The current divides itself 'automatically' into its sinusoidal components. This effect can be used, on the one hand, as a method of harmonic measurement or, on the other hand, as will be seen later, to filter out unwanted harmonics.

Only the fundamental components of the voltage and current contribute to power translation. The harmonics only load the translation system and cause an increase in the apparent power in the source. The non-active power due to the harmonics may be referred to as *distortion power* (distortion volt-amperes).

Fourier analysis makes calculations in circuits with non-sinusoidal currents and voltages easier. If a waveform of any degree of complexity is resolved into its sinusoidal components, calculation can be carried out much more simply with the individual sine waves and the required result obtained by synthesis.

PRINCIPLES OF HARMONICS

The alternating current in the three-phase bridge converter does not have a sinusoidal waveform (with a few exceptions) but, depending upon the loading and the control, assumes waveforms very different from the sine wave. In the ideal case (smooth direct current, no commutating inductance) the waveform is rectilinear (Figure 72). Commutation makes the current waveform more trapezoidal (Figure 75).

Fourier analysis of the ideal rectilinear wave shows that besides the fundamental component (I_1) it contains harmonics of orders $n = 5, 7, 11, 13, 17, 19, 23, 25$, etc. (Figure 85). Even when the current is distorted as a result of commutation, ripple, or the nature of the load, only the amplitudes and phase angles of the harmonics are altered, not their orders.

It can be shown that in the three-phase bridge under steady-state conditions the alternating current can contain only harmonics of orders $n = 6 \pm 1$, 12 ± 1, 18 ± 1, etc. In the ideal rectilinear waveform the amplitudes of the individual harmonics are inversely proportional to their orders n. The resulting amplitude spectrum is shown in Figure 85. *The higher the order n of the harmonic, the lower is its amplitude.* Figure 86 shows the effects of ripple and the trapezoidal shape on the amplitudes of the first three harmonics. With increasing direct current ripple, the fifth harmonic increases, the seventh decreases, and the eleventh remains practically constant. In the trapezoidal waveform the overlap μ diminishes all the harmonics.

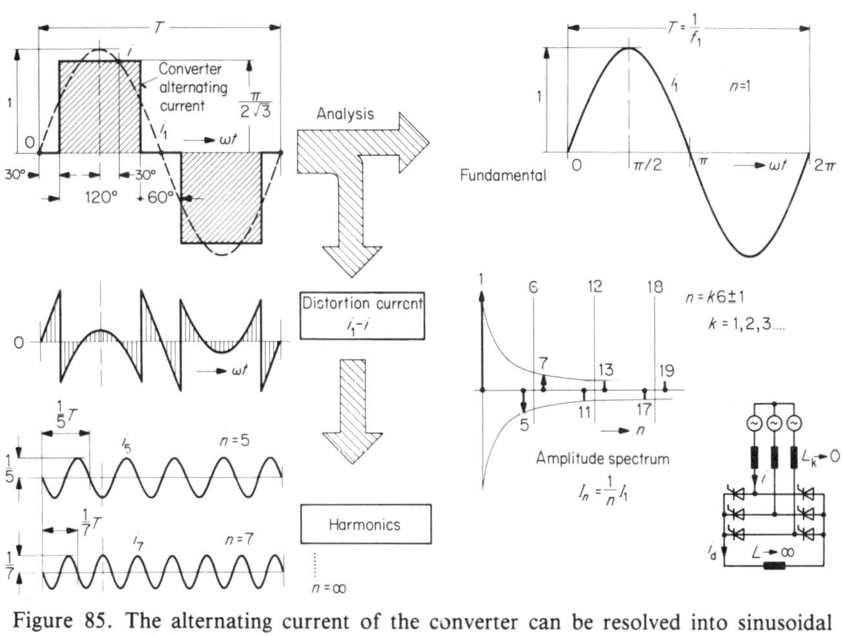

Figure 85. The alternating current of the converter can be resolved into sinusoidal components of the same and higher frequencies by means of Fourier analysis. These are referred to as the fundamental and the harmonics. Example of the analysis of the ideal rectilinear wave current. Waveforms and spectrum

In Figure 86 the harmonics I_n are scaled in relative values. The reference quantity is the actual fundamental component I_1. The current waveforms in Figure 86 are considerably idealized, but they illustrate the harmonic tendencies correctly. In Figure 87 the Fourier analysis of the alternating current is again shown for a smooth direct current (for $n = 5$ and 7), but in this case the effect of commutation μ on the waveform is taken accurately into account. In addition, the range of amplitude variation due to delay angle is indicated ($0° \leqslant \alpha \leqslant 90°$). With increasing commutation angle μ and constant delay angle α the current I_d increases (see Figure 74). The relative harmonics decrease and the current more nearly approaches the sinusoidal.

Fourier analysis of the *direct voltage* v_d of the converter (Figures 72 and 75, top) leads to the result that *the three-phase bridge produces harmonics only of the orders* $n = 6$, 12, 18, etc. Figure 87 (bottom) shows how the sixth direct

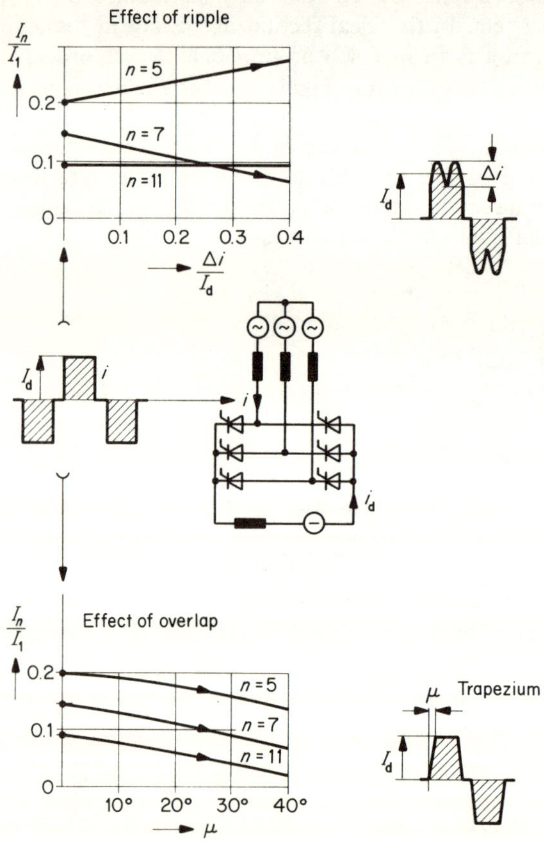

Figure 86. Departure from the ideal rectilinear waveform of the alternating current changes the harmonic amplitudes, but the orders $n = 5$, 7, 11, 13, etc., remain fixed. Idealized current waveforms: top, sinusoidal current fluctuations Δi; bottom, trapezoidal wave current due to commutation

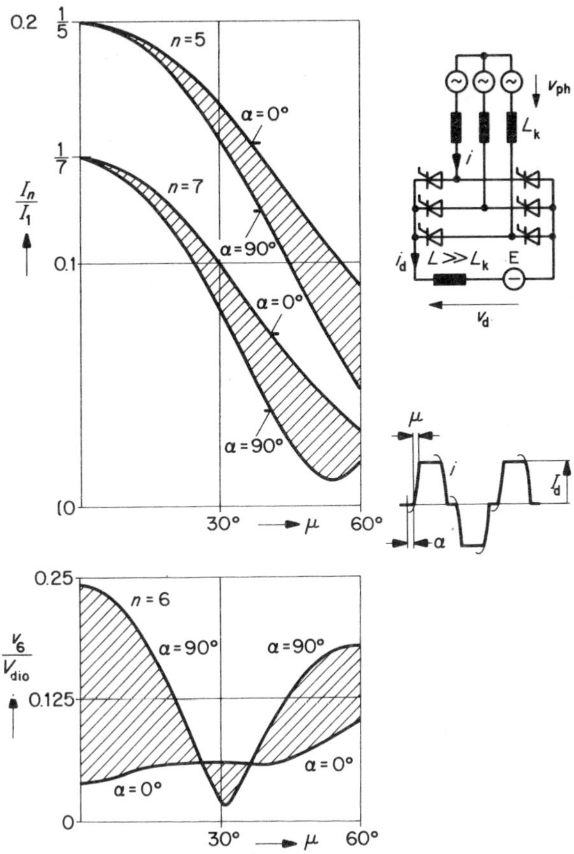

Figure 87. The fifth and seventh harmonics of the alternating current decrease with increasing overlap μ (increasing current i_d)(top). The delay angle α had relatively little effect. The sixth harmonic in the direct voltage, on the other hand (Figures 41 and 75), varies considerably with the delay angle α (bottom)

voltage harmonic varies in amplitude with the overlap angle μ over the whole range of delay angles $0^\circ \leqslant \alpha \leqslant 90^\circ$ (corresponding, in the inverter mode, to $180^\circ \geqslant \alpha \geqslant 90^\circ$). Figure 88 (bottom) shows the variation of the amplitudes of the fifth and seventh alternating current harmonics in the diode bridge without load inductance. In this case the harmonics are related to the short-circuit current I_{dk}. As well as the a.c.-side harmonics, the variation of the sixth harmonic in the direct current is also shown here.

The harmonics in the three-phase bridge with *purely inductive load,* relative to the actual fundamental component, can be seen in Figure 89. In the discontinuous current region ($90^\circ \leqslant \alpha \leqslant 120^\circ$) the harmonics increase sharply with increasing delay angle α. In the range $0^\circ < \alpha < 30^\circ$ operation is not stable (Figure 53e), and the ultimate condition of the converter corresponds to

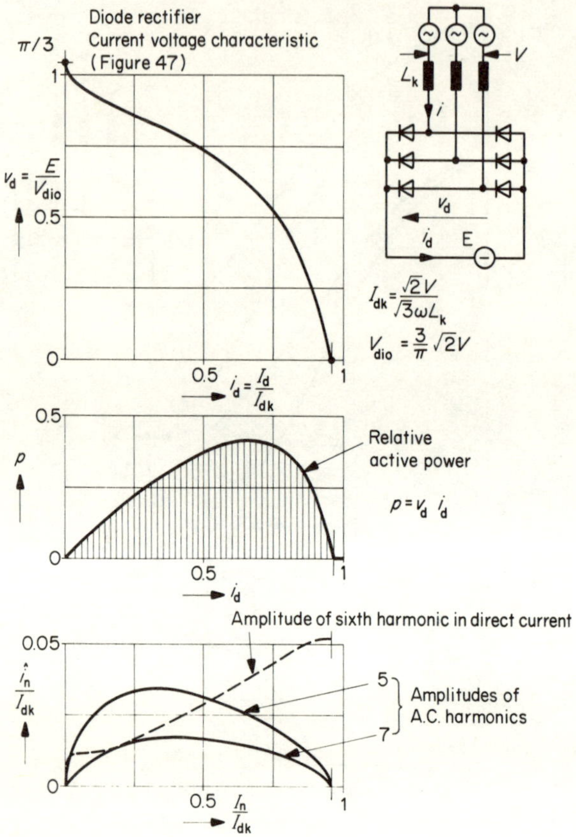

Figure 88. In a diode bridge with no load inductance the sixth harmonic in the direct current of the converter increases continuously with increasing current i_d. The alternating current harmonics ($n = 5, 7$), on the other hand, like the active power p, go through a maximum

$\alpha = 60^\circ$. With a smaller inductance L the fifth harmonic is greater, and the seventh less, than with a larger L (the same tendency as in Figure 86, top).

The alternating current harmonics with *purely resistive load* are shown in Figure 90 as a function of the delay angle α. With increasing delay angle the relative harmonics (related to the actual fundamental component) increase steeply (right-hand graph). However, since the absolute value of the fundamental component I_1 decreases at the same time (left-hand graph), the absolute values of the harmonics hardly change. (The load resistance R is constant.) Only the proportion of harmonics k_H increases. The sixth harmonic in the direct voltage v_d (and in the direct current i_d) goes through a maximum (Figure 90, top).

The effect of the load inductance L on the alternating current harmonics ($n = 5$, 7, 11, and 13) for a particular example ($\alpha = 30^\circ$, $E/V_{dio} = 0.8$) is illustrated in Figure 91. It will be seen that the general tendency is confirmed:

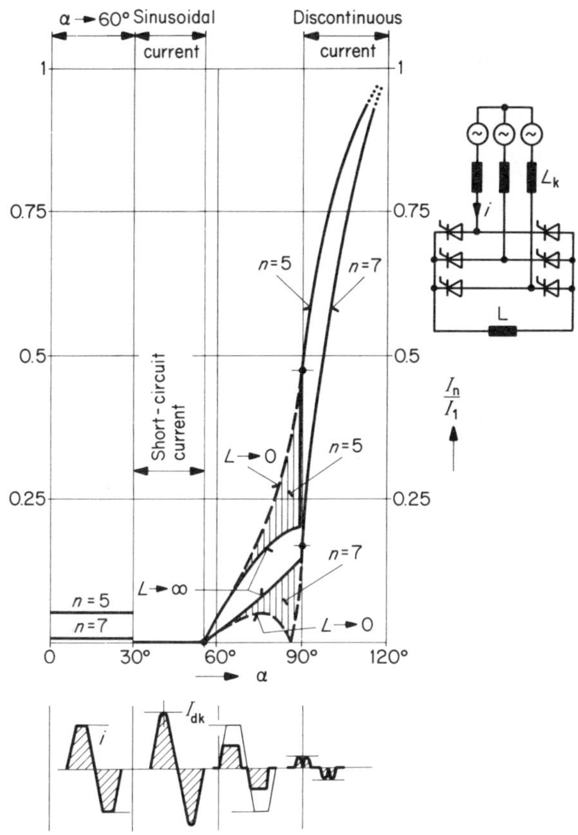

Figure 89. The relative harmonic current ($n = 5, 7$) in the three-phase bridge with purely inductive load increases steeply with increasing delay angle α, particularly in the discontinuous current region. With $0° < \alpha < 30°$ a stable condition is not possible (Figure 52). With $30° < \alpha < 55°$ the alternating current is sinusoidal (Figures 53 and 59)

increasing ripple leads to an increase in the fifth harmonic and a decrease in the seventh (Figure 86, top). At the top of the figure the current waveforms are shown for three different inductance ratios ($L/L_k = 0.6$, 6, and 60). The amplitude spectra of the alternating and direct current and of the alternating and direct voltages are shown in Figure 92 for one case ($L/L_k = 6$). It will be observed that *for each d.c.-side harmonic there are two corresponding a.c.-side harmonics.* The variation of the harmonic amplitudes with load current at a constant delay angle ($\alpha = 30°$) is shown in Figure 93. The alternating current harmonics are here given in both absolute and relative values (bottom left)(V = source voltage).

To summarize briefly: *the three-phase bridge converter supplies the load with a six-pulse waveform,* and harmonics of order six and multiples of six are therefore present in the load voltage. The same harmonics are contained in the load current. *In the alternating current, on the other hand, harmonic*

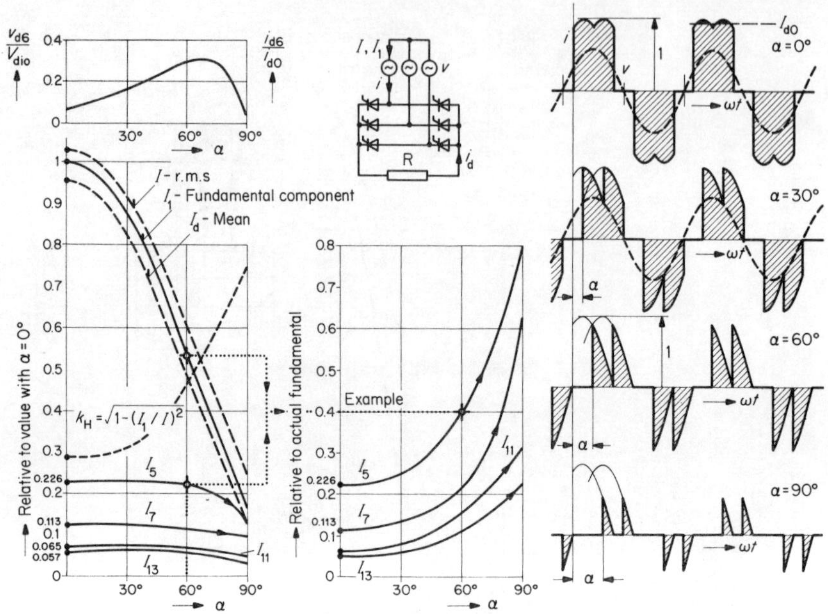

Figure 90. In an ideal converter with resistive load the harmonic amplitudes are practically unaffected, in absolute terms, by the delay angle. On the other hand, the relative values (right) and the harmonic content k_H increase considerably with increasing α. The direct current harmonic ($n = 6$) goes through a maximum (top)

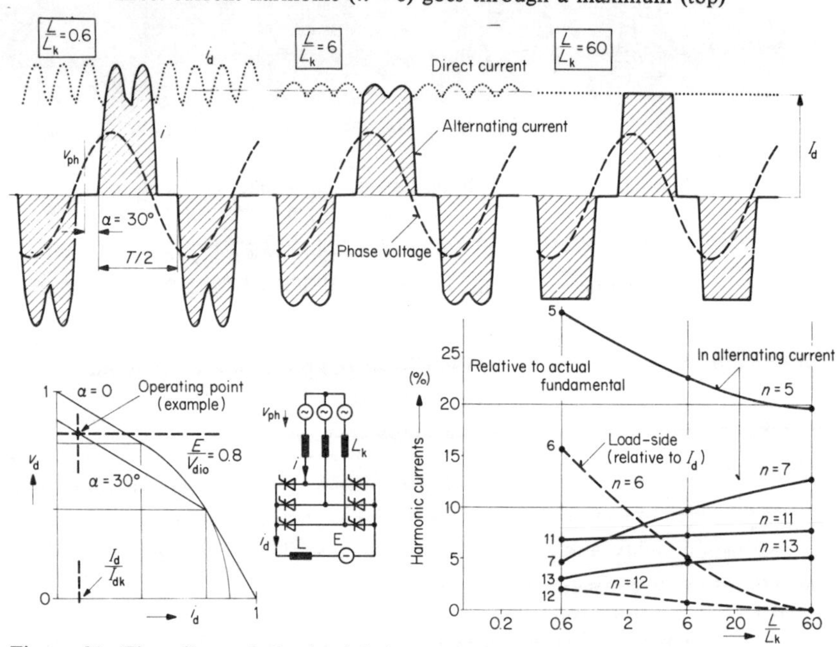

Figure 91. The effect of the load inductance L on the harmonics is particularly noticeable on the d.c. side ($n = 6$, 12). The fifth harmonic in the alternating current decreases with increasing L, the seventh increases (Figure 86, top). Example with $\alpha = 30°$, $E/V_{dio} = 0.8$, $L/L_k = 0.6$, 6, 60

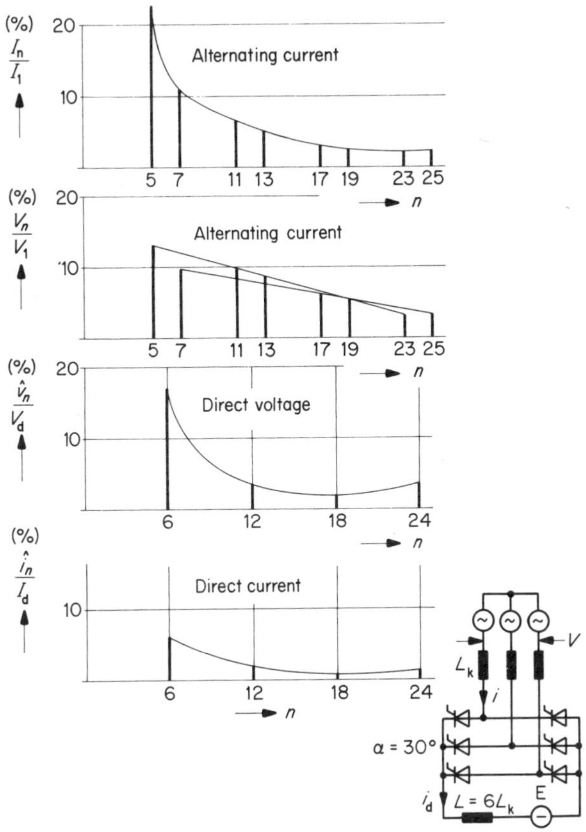

Figure 92. The alternating current and voltage in the three-phase bridge contain harmonics of orders $n = k6 \pm 1$ ($k = 1, 2, 3, \ldots$). On the load side appear harmonics of orders $n = k6$. Example of amplitude spectra with $\alpha = 30°$, $L/L_k = 6$, $E/V_{dio} = 0.8$ (Figure 91)

components appear with $n = 6 \pm 1$, 12 ± 1, 18 ± 1, etc. The converter can therefore usefully be regarded as a *harmonic generator*. In the three-phase system the converter functions as a series of current sources producing the harmonics $n = 5, 7, 11, 13, 17, 19, 23, 25$, etc. On the d.c. side the converter functions as a series of harmonic voltage sources with $n = 6, 12, 18, 24$, etc. Figure 94 shows the relative equivalent circuit.

In the study of supply loading the converter can usually be regarded as a harmonic current source in accordance with the equivalent circuit of Figure 94. There are, however, limitations to this view. The supply network parameters are of similar significance to the load parameters and the delay angle in determining the current waveform and therefore the harmonic spectrum. If the supply inductance is a small proportion of the total commutating inductance, a variation in the supply inductance produces little variation in the harmonic spectrum and the converter can be regarded as a current source. Otherwise the

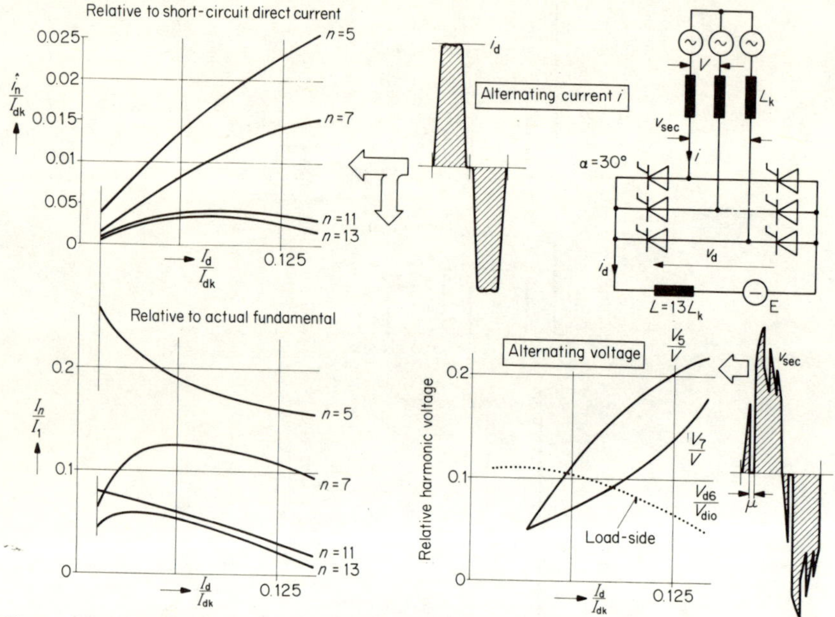

Figure 93. The effect of current loading I_d on the harmonics in the three-phase bridge. The relationships appear different according to the reference quantity. Top: absolute values. Bottom left: relative values. Example with $\alpha = 30\,°$, $L/L_k = 13$

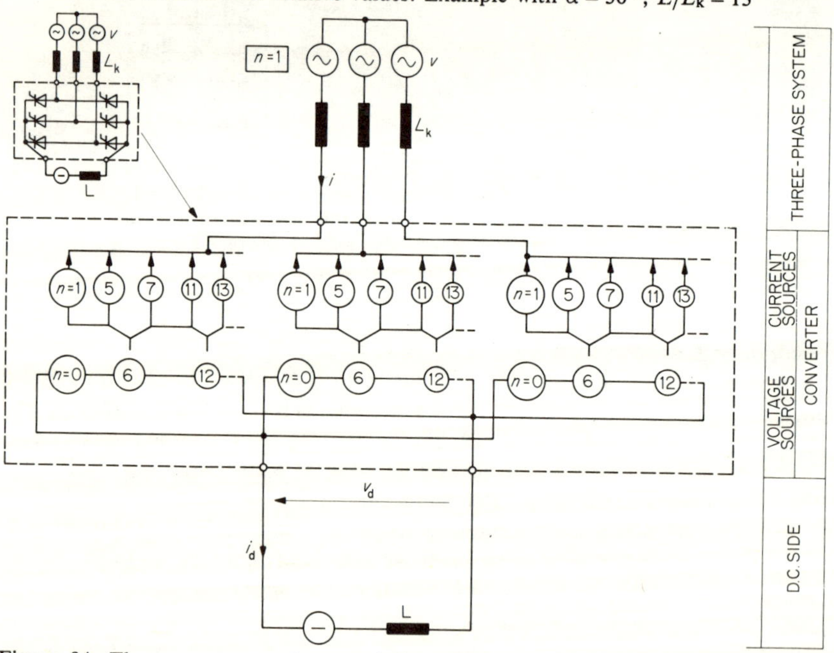

Figure 94. The converter can be regarded as a harmonic generator. On the a.c. side the three-phase bridge functions as a series of current sources with orders $n = k6 \pm 1$. The harmonics with $n = k6 + 1$ have positive phase sequence (relative to the fundamental), those with $n = k6 - 1$ negative. On the d.c. side voltage sources appear with orders $n = k6$

supply network affects the current waveform and the converter no longer behaves as an ideal current source. This is especially so when the supply network includes capacitance, as will be shown later.

TRANSIENT CONDITIONS

In the steady-state condition the alternating current in the three-phase bridge is distorted and non-sinusoidal, but it is symmetrical and its fundamental frequency is the same as that of the alternating voltage source. The harmonic spectrum in this case contains the orders $n = k6 \pm 1$ for all operating conditions and any kind of loading.

If the current is asymmetrical, harmonics of other orders are introduced, as illustrated in Figure 95. (Slight asymmetries can occur in practice under steady-state conditions as a result of unequal commutating inductances or inexact control.)

During the build-up of converter current with constant delay angle α,

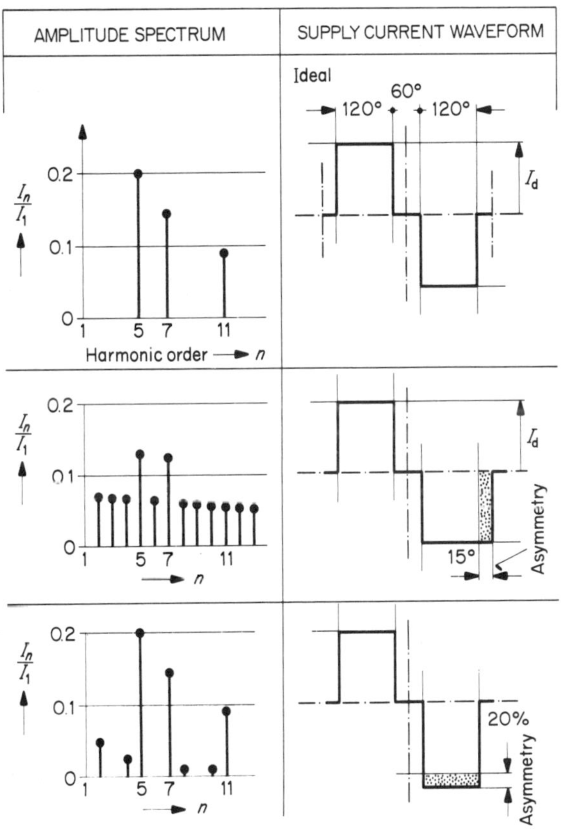

Figure 95. The ideal rectilinear wave alternating current of the three-phase bridge (top) contains harmonics of orders $n = k6 \pm 1$ with amplitudes that decrease according to the law $I_n = I_1/n$. Other harmonics arise through asymmetries: $n = 2, 3, 4, \ldots$ (middle, bottom). Example

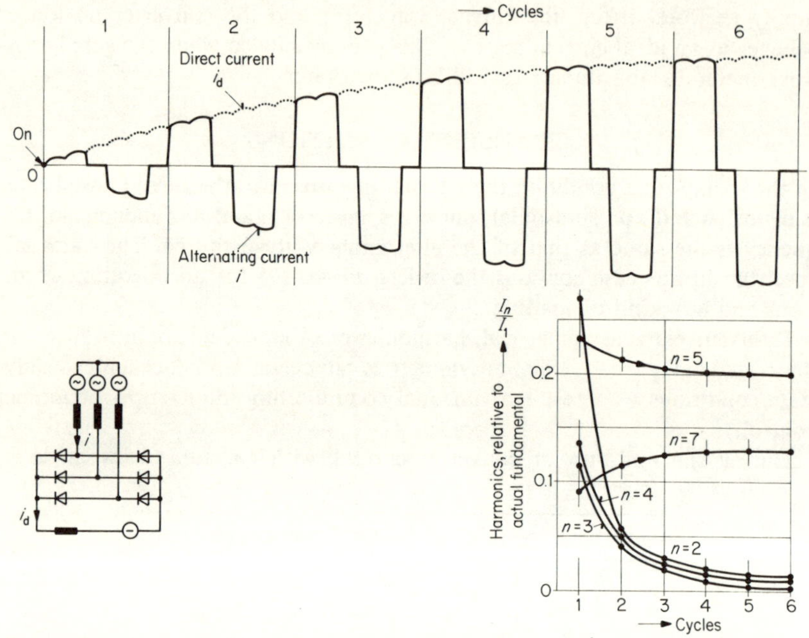

Figure 96. Current build-up in a converter with constant back-voltage and constant delay angle α. During the transient period, in addition to the characteristic harmonics ($n = 5, 7, \ldots$), harmonics of other orders appear ($n = 2, 4$). Example

besides the characteristic harmonics $n = k6 \pm 1$, second, third, and fourth harmonics are introduced. Figure 96 shows the way in which the individual harmonics vary in particular cycles during the rise of current. When the steady-state condition is reached only the characteristic harmonics—the fifth, seventh, etc.—remain.

In contrast to Figure 96, where the current build-up is with constant back-voltage E and constant delay angle α, Figure 97 illustrates acceleration with constant direct current and linearly increasing back-voltage (see Figure 83). Here also, in the transient state, harmonics other than those of the characteristic orders occur.

MATHEMATICAL APPENDIX XIV

The ideal rectilinear wave alternating current in the three phase bridge (Figure 85) contains harmonics of the following orders:

$$n = k6 \pm 1 \qquad \text{for } k = 1,2,3,\ldots \tag{90}$$

i.e. $n = 5, 7, 11, 13, \ldots$

The amplitudes of the harmonics I_n are inversely proportional to the order n:

$$I_n = \frac{1}{n} I_1 \tag{91}$$

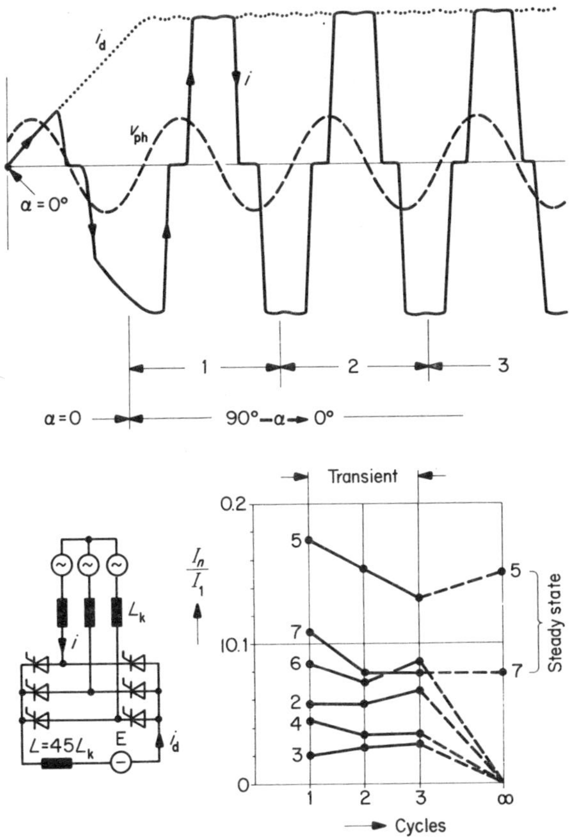

Figure 97. Build-up in a converter with constant current and variable delay angle α. The back-voltage rises linearly (Figure 83). During the transient period harmonics of all orders appear. Under steady-state conditions only the characteristic harmonics remain. Example

i.e. $I_5 = \frac{1}{5} I_1$, $I_7 = \frac{1}{7} I_1$, etc.

The more harmonics are taken into account in the analysis (or synthesis), the more accurately is the current waveform represented by the sum of the sinusoidal components. Figure 98 shows how the rectilinear waveform is approximated by seven components ($n = 1, 5, \ldots, 19$).

The r.m.s. value of the current is

$$I = \sqrt{\frac{2}{3}}\, I_d = 0.8165 I_d \tag{92}$$

and the fundamental component is given by (Figure 58)

$$I_1 = \frac{3}{\pi} I = 0.955 I \tag{93}$$

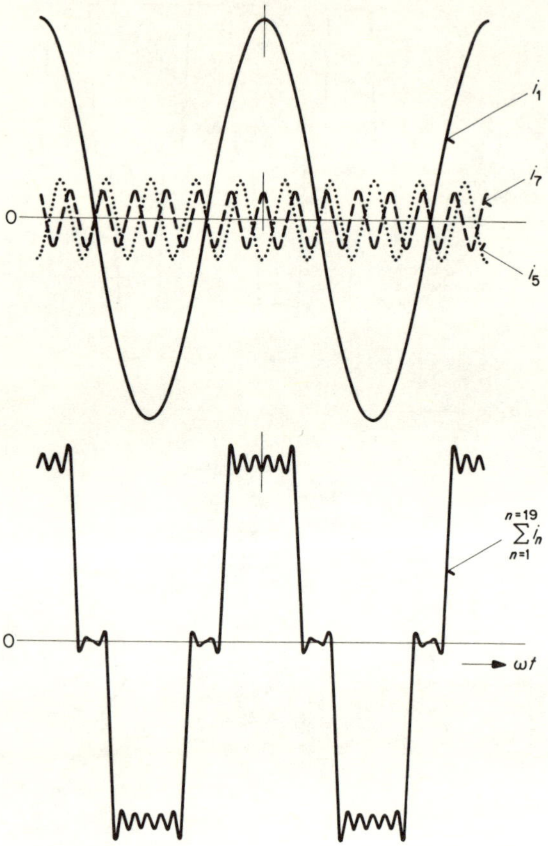

Figure 98. The greater the number of harmonics taken into account in Fourier analysis (or synthesis), the more closely is the analysed waveform approximated. Top: fundamental component i_1 and the fifth ($i_5 = i/5$) and seventh ($i_7 = i/7$) harmonics in the rectilinear waveform (Figure 85). Bottom: sum of $n = 1$ to 19

The fundamental content is therefore

$$k_1 = \frac{I_1}{I} = \frac{3}{\pi} = 0.955 \tag{94}$$

and the harmonic content is given by

$$k_H = \sqrt{1 - k_1^2} = 0.297 \tag{95}$$

The ideal alternating current in the three-phase bridge maximum output $\alpha = 0°$ with resistive load (Figure 90) has an r.m.s. value of

$$I = \frac{\pi}{3\sqrt{3}} \sqrt{1 + \frac{3\sqrt{3}}{2\pi}} = 0.8172 I_d \tag{96}$$

The fundamental component is

$$I_1 = 0.781 I_d \tag{97}$$

The fundamental content is

$$k_1 = \frac{I_1}{I} = \frac{0.781}{0.8172} = 0.956 \tag{98}$$

and the harmonic content is

$$k_H = \sqrt{1 - k_1^2} = 0.294 \tag{99}$$

For the first two harmonics,

$$I_5 = 0.226 I_1 \qquad \text{and} \qquad I_7 = 0.113 I_1 \tag{100}$$

In the direct voltage harmonics are present with orders

$$n = k6 \qquad \text{for } k = 1, 2, 3, \ldots \tag{101}$$

i.e. $n = 6, 12, 18, \ldots$

The r.m.s. value of the harmonics in the ideal direct voltage (without commutation; see Figures 51, 60 (top), and 72) can be calculated as

$$V_n = V_{dio} \frac{\sqrt{2}\sqrt{n^2 - (n^2 - 1)\cos^2\alpha}}{n^2 - 1} \tag{102}$$

10

Fault Conditions in the Three-Phase Bridge

GENERAL

By a fault condition is implied, as distinct from the normal condition, an unintended modification of the switching pattern or the control of the converter.

Breakdown of a thyristor: the valve conducts permanently in both directions.

Blocking of a thyristor: the valve is in a permanent blocking state.

In the control system, as a result of a fault, the *triggering pulses* can occur at unintended times or be directed to the wrong thyristors.

On the load side, a *short circuit* can occur and finally there can be a failure of distortion of the *alternating voltage supply*.

All these fault conditions can arise in either the rectifier or the inverter mode of the converter.

The fault conditions referred to above in the three-phase bridge are discussed below without, however, considering the appropriate protection measures (for these, see chapter 14 on converter protection).

The probability of occurrence of particular faults will not be considered initially, since this question belongs to the protection philosophy of the converter. It will also be assumed that there is only one valve in each branch of the bridge (no parallel or series connection).

VALVE BREAKDOWN

One of the commonest faults in converters with semiconductor valves, thyristors or diodes, is the sudden loss of the forward or reverse blocking capability of a valve. In this event the valve conducts continuously. In the equivalent circuit the valve thus behaves subsequently like a conductor.

The failed valve produces an *internal short-circuit current* in the converter. Figures 99 and 100 show the unmodified short-circuit current for three different cases ($\alpha = 0°$, $30°$, and $150°$) on the assumption that the other thyristors continue to be controlled normally and that no protection measures operate. In the maximum-output condition ($\alpha = 0°$, Figure 99) the failed thyristor carries a sinusoidally modulated direct current with no zero crossings, with a mean value of $\sqrt{3}\, I_{dk}$. (In a diode rectifier it reaches $2I_{dk}$.) In this condition the circuit behaves as a two-pulse converter (Figure 99, bottom). With a delay

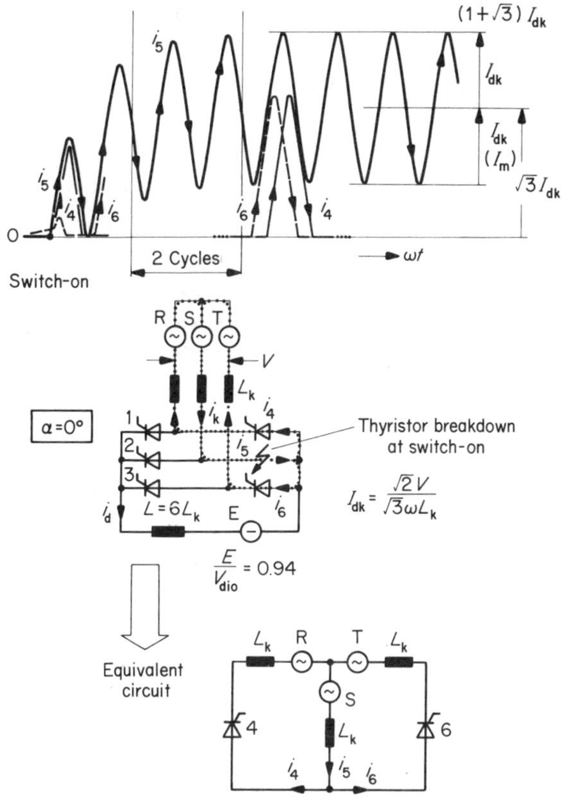

Figure 99. Valve breakdown in the three-phase bridge leads to an internal short-circuit current which far exceeds the limit I_{dk} of the converter characteristic (Figures 43, 44, 56). The circuit functions in this event as a two-pulse converter. Example with $\alpha = 0°$. The thyristor (5) fails at switch-on

angle greater than 30° the failed thyristor carries a discontinuous direct current (Figure 100, top). If the converter is operating in the inverter mode, the valve failure causes a *loss of control* (Figure 100, bottom). The failed thyristor, together with the thyristor in the same phase in the other commutating group, presents a low-impedance path for short-circuit current supplied by the d.c. voltage source.

INOPERATIVE VALVE

If a thyristor in the three-phase bridge fails to trigger and therefore cannot conduct (or if the valve is disconnected by a fuse), the converter is placed in an unstable condition. Figure 101 shows the direct current i_d and the direct voltage v_d in the rectifier mode, as an example, with $\alpha = 30°$. The rectifier in this case is periodically *in a switching-on condition.*

In the inverter mode an inoperative valve leads to a *loss of control.* When

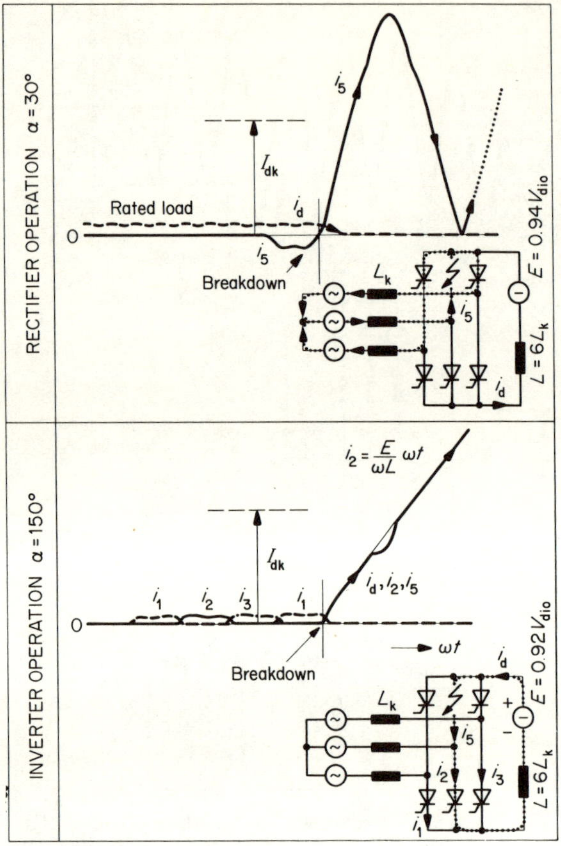

Figure 100. Valve breakdown in rectifier and inverter operation. Example with $\alpha = 30°$ and $\alpha = 150°$. Comparison with Figure 99 (scale changed). The inverter loses control following the establishment of a by-pass path through the failed thyristor. Breakdown in normal operation

the next thyristor in the switching sequence fails to trigger, the thyristor that is carrying current continues to conduct and the circuit develops a short circuit. This is illustrated in Figure 102.

PHASE FAILURE

If one phase of the three-phase system is lost, the three-phase bridge can continue to operate as a *two-pulse converter*. With diodes (Figure 27), the circuit assumes the normal load characteristic of the two-pulse converter, as shown in Figure 103 (for $L \gg L_k$) and Figure 104 (for $L = 0$).

In a controlled three-phase bridge the pulse positions in the thyristor control system are not exactly those required to correspond to the single-phase bridge. The converter cannot therefore operate at maximum output in the two-pulse mode and under certain conditions exhibits freewheeling behaviour. Figure

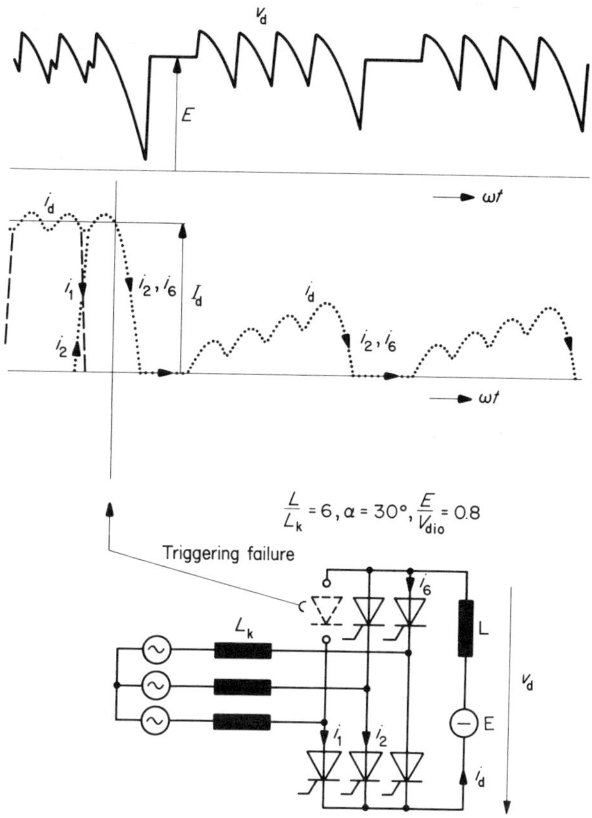

Figure 101. With a non-conducting valve (triggering pulse failure, fuse rupture), the operation of the three-phase bridge is greatly impeded. Example with $\alpha = 30\,°$, $L/L_k = 0.6$ and $E/V_{dio} = 0.8$ (see Figure 91, middle). The switch-on process is repeated periodically. Rated load

103 shows the load characteristic for $\alpha = 0°$ in the two-pulse mode (relative to three-phase bridge operation).

Figure 104 shows the load characteristic of the three-phase bridge operating in the single-pulse mode ($L = 0$).

D.C.-SIDE SHORT CIRCUIT

The inductance L and the load back-voltage E in the equivalent circuit of the three-phase bridge simulate a d.c. machine in a d.c. drive. Possible fault conditions include *commutator flash-over, terminal short-circuit on the machine, terminal short circuit on the converter,* and *full-output drive* with the machine at standstill. Under all these fault conditions, it can be assumed that $E = 0$ and that either the circuit includes a certain inductance L or, in an extreme case, that $L = 0$. The fault can occur with the converter initially loaded or unloaded. If the d.c. circuit retains a large inductance $L \gg L_k$, the steady-state short-

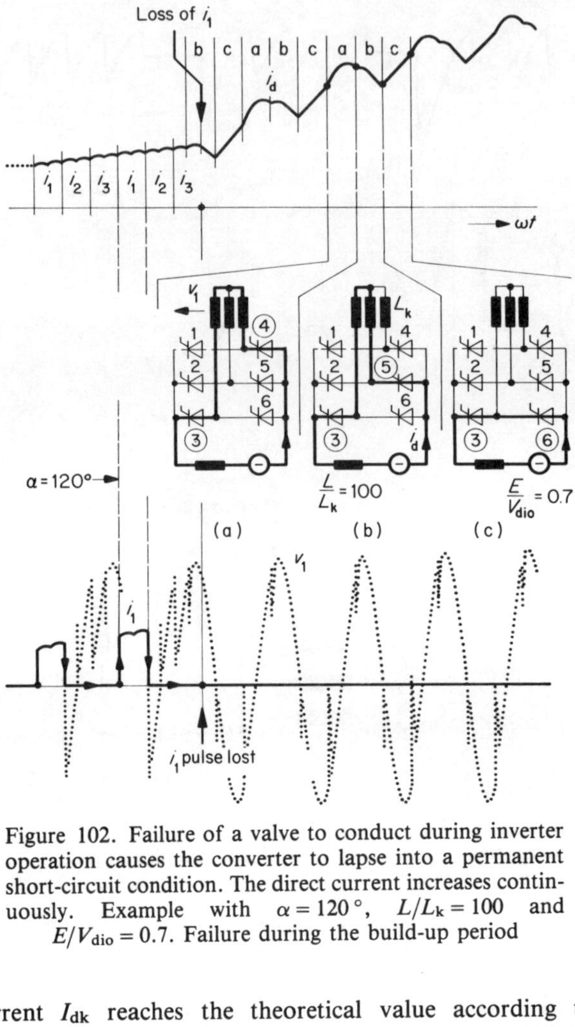

Figure 102. Failure of a valve to conduct during inverter operation causes the converter to lapse into a permanent short-circuit condition. The direct current increases continuously. Example with $\alpha = 120°$, $L/L_K = 100$ and $E/V_{dio} = 0.7$. Failure during the build-up period

circuit current I_{dk} reaches the theoretical value according to the load characteristics (Figures 43 to 46). This shown (for $L = 6L_K$) in Figure 105. At a maximum the direct current in the short-circuit condition reaches a level equal to the amplitude of the alternating current corresponding to a symmetrical three-phase short circuit on the supply. The converter in this case contains either diodes or thyristors triggered at $\alpha = 30°$. If α is reduced to $0°$ (narrow double pulses; see Figures 48 and 49) the converter jumps to the characteristic for $\alpha = 60°$ (Figures 52 and 53) and the maximum short-circuit current is reduced. This is shown in Figure 106 (bottom).

With a small load inductance ($L \ll L_K$), on the other hand, the short-circuit current has a different waveform and rises to a higher value than with a large load inductance. This is shown in Figure 107. (The dependence of the d.c. short-circuit current upon the inductance ratio L/L_K is shown in Figure 106, top). In the extreme case $L = 0$, no freewheeling can take place and the direct

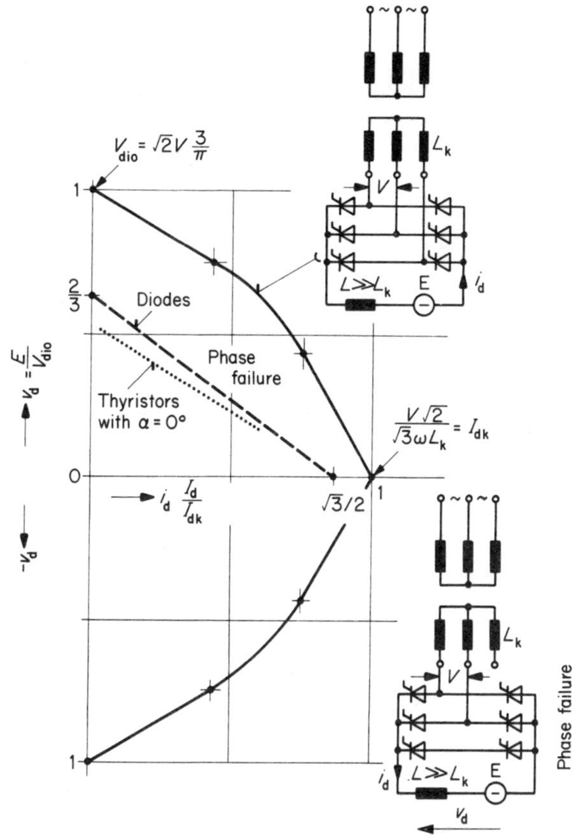

Figure 103. Upon failure of one phase of the three-phase system the three-phase bridge continues to function in a two-pulse mode (over a limited range). Steady-state characteristics for diode operation and for thyristors with triggering pulses at $\alpha = 0°$, with high load inductance (relative to three-phase bridge operation)

current is discontinuous (Figure 107, for $\alpha = 0°$ at the limit of discontinuity). The maximum value is $1.87I_{dk}$.

In the diode bridge the direct current in the case of a short circuit can reach a maximum value of $2I_{dk}$.

A d.c.-side short circuit in inverter operation unloads the converter. The direct current is either zero, with $\alpha > 120°$, or discontinuous, with $90° < \alpha < 120°$ (see Figures 64, 66, and 67).

SUPPLY FAILURE

In a d.c. drive the three-phase system is provided by the supply mains. A mains voltage dip or failure during rectifier operation of the three-phase bridge

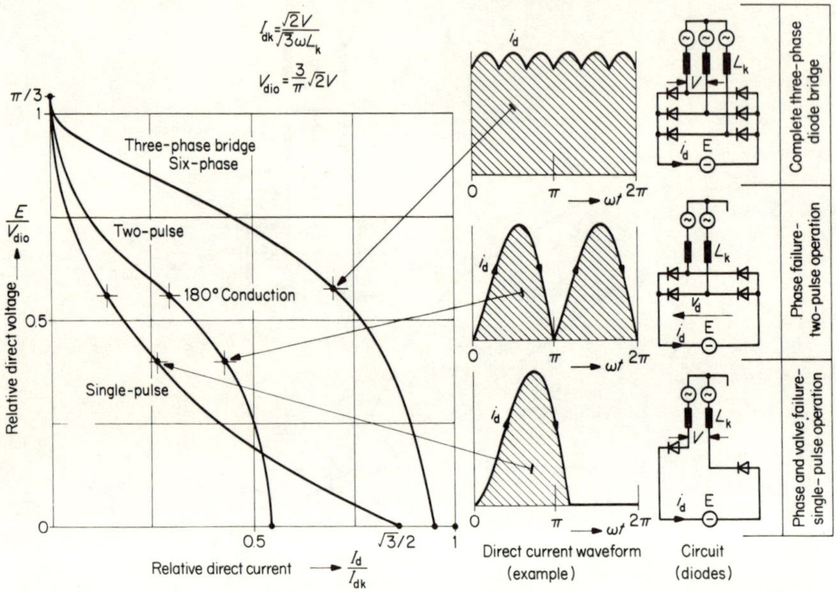

Figure 104. Characteristics of the three-phase bridge with very low load inductance, $L \ll L_k$, in diode operation. Normal operation and two fault situations: phase failure and phase failure with additional inoperative valve. Steady-state characteristics. For high inductance see Figure 103

results in a proportional drop in the direct voltage. In the relative characteristic diagram (Figure 56) the drop in alternating voltage V is equivalent to an increase in the direct voltage E. The direct current therefore becomes smaller, or even zero (see Figure 64).

In inverter operation, on the other hand, a reduction in alternating voltage causes an increase in current, which can lead to a loss of control. This can be seen in Figures 108, 109, and 110.

As a result of the supply voltage dip (to 50 per cent of the nominal value in all phases) the current in Figure 108 rises and the overlap μ becomes progressively larger (E and α remaining constant). At the critical instant the next thyristor in sequence (3) is triggered, although commutation between thyristors 6 and 4 is not complete. Thyristor 4 does not therefore take over the current and turns off. Thyristors 6 and 3 then remain permanently conducting (in a by-pass path). Commutation has failed and the inverter is no longer controllable.

Figure 109 shows a similar case to a different time scale. The by-pass path in this case is through thyristors 1 and 4. The direct current i_d increases linearly.

A different mode of inverter failure is illustrated in Figure 110. Here the short-circuit path is through two commutating inductances, a.c. voltage sources, load inductance, and d.c. voltage source. The direct current is therefore modulated by the alternating voltage. Thyristors 1 and 6 conduct continuously.

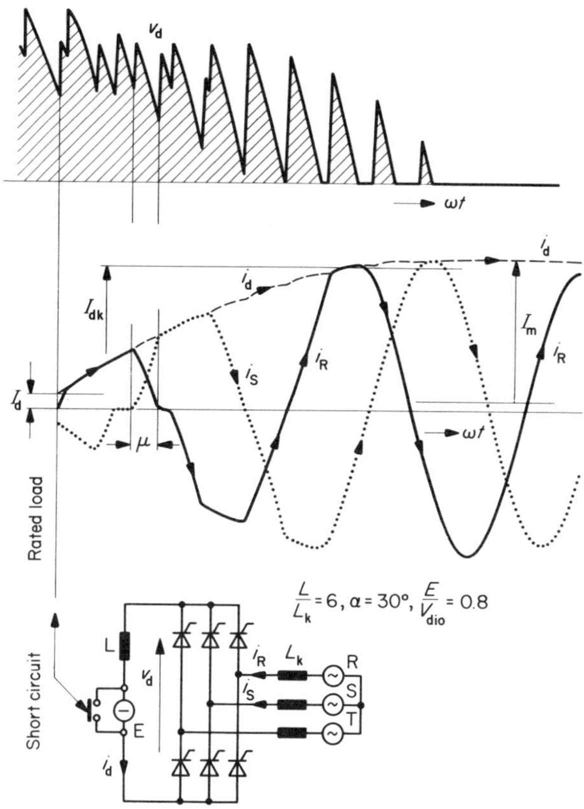

Figure 105. In the event of a short circuit the direct current i_d of the three-phase bridge attains a maximum value equal to the amplitude of the short-circuit alternating current. Example with $\alpha = 30°$ and $L/L_k = 6$. Initial load with $E/V_{dio} = 0.8$. Compare with Figure 106. Rectifier operation

ESTIMATION OF FAULT CONDITIONS

Figurc 111 indicatcs diagrammatically the short-circuit currents that result from particular fault conditions. These are derived from the load diagram for the three-phase bridge.

The short-circuit current due to inverter failure is theoretically unlimited. The short-circuit current rises relatively slowly, however. The currents in the case of a valve breakdown can reach very high levels in a short time. The most severe loading is produced by an already failed valve.

MATHEMATICAL APPENDIX XV

The reference value adopted for the steady-state load characteristics of the three-phase bridge is the d.c.-side short-circuit current, whose mean value I_{dk}

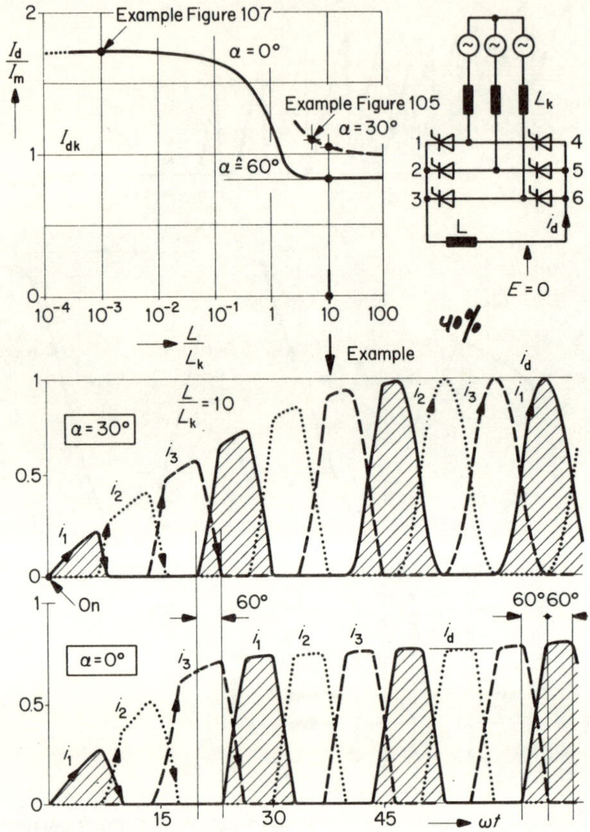

Figure 106. Relationship of the short-circuit current of the three-phase bridge with minimum delay angle to the inductance ratio (top) ($L/L_k = 6$; see Figure 105). For waveforms with very low inductance see Figure 107. Bottom: example with $L/L_k = 10$, $\alpha = 0°$ and 30°

is equal to the peak symmetrical a.c. short-circuit current I_m:

$$I_{dk} = I_m = \frac{\sqrt{2}\,V}{\sqrt{3}\,\omega L_k} \tag{103}$$

With a small load inductance ($L \ll L_k$) the d.c.–side short-circuit current reaches a higher value, however (Figure 107).

In the event of *inverter failure* (Figures 108 to 110) the short-circuit current is supplied by the d.c. voltage source through the circuit inductance (L or $L + 2L_k$) and therefore increases linearly and indefinitely:

$$i_{dk} = \frac{E}{\omega L}\,\omega t \tag{104}$$

In practice the maximum value is limited by the internal resistance of the d.c. voltage source (the d.c. machine, for example).

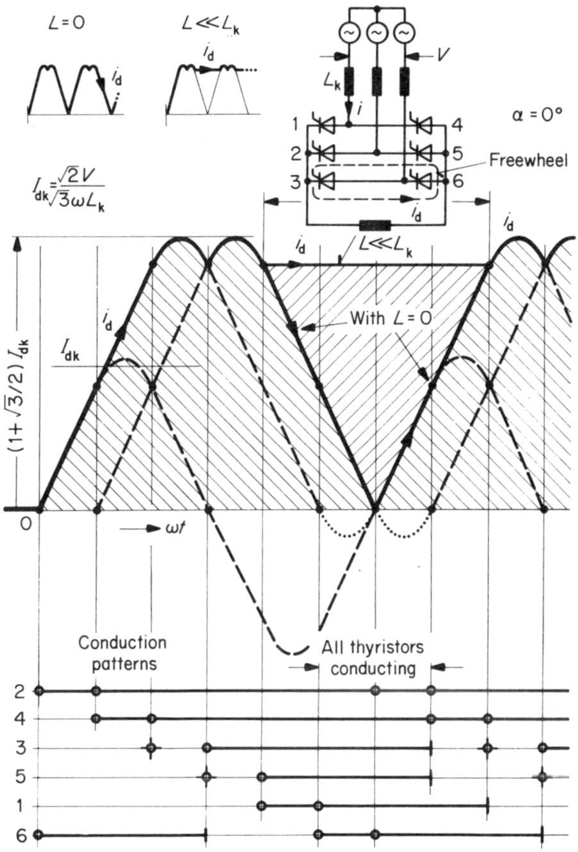

Figure 107. With a perfect short circuit ($L=0$, $E=0$) on the d.c. terminals of the three-phase bridge the direct current with $\alpha = 0°$ pulses with an amplitude equal to $(1+\sqrt{\frac{3}{2}})I_{dk}$. The mean value is $1.18 I_{dk}$ and the r.m.s. value is $1.33 I_{dk}$. The fundamental component is $0.84 I_{dk}$. With minimal inductance L freewheeling takes place and the current is smooth. Rectifier operation. $\alpha = 0°$

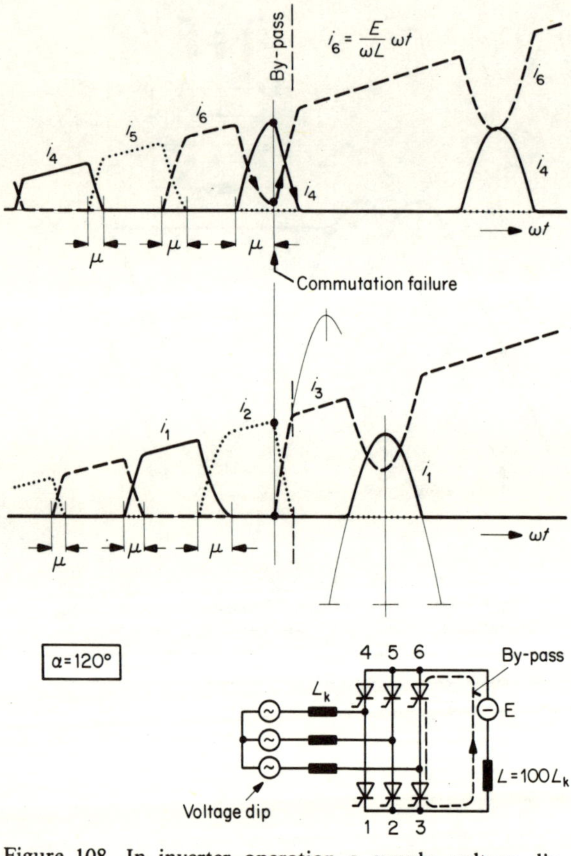

Figure 108. In inverter operation a supply–voltage dip causes loss of control. Following an unsuccessful commutation the direct current increases indefinitely. The short-circuit current flows through two thyristors, which form a by-pass. Example with $\alpha = 120°$

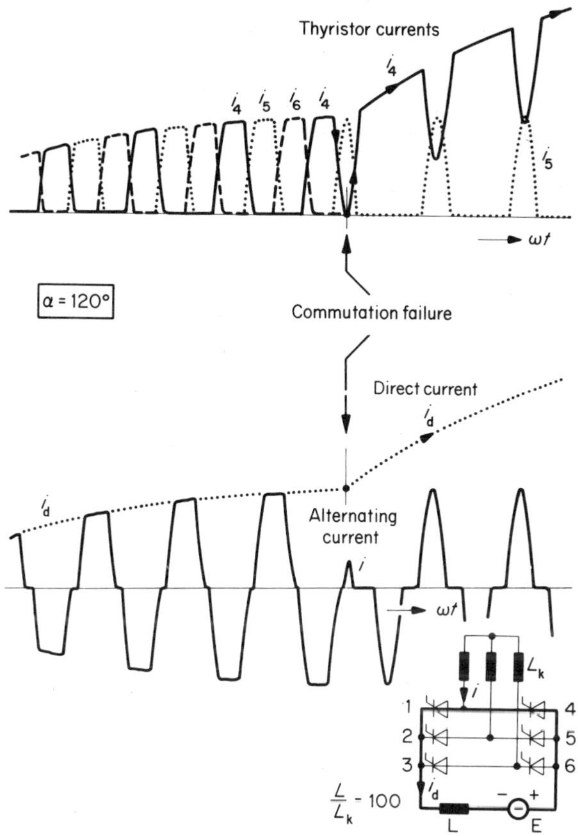

Figure 109. Loss of inverter commutation as in Figure 108. Continuous increase in direct current through two thyristors forming a by-pass. The converter can no longer be influenced by control. Example with $\alpha = 120°$ and $L/L_k = 100$

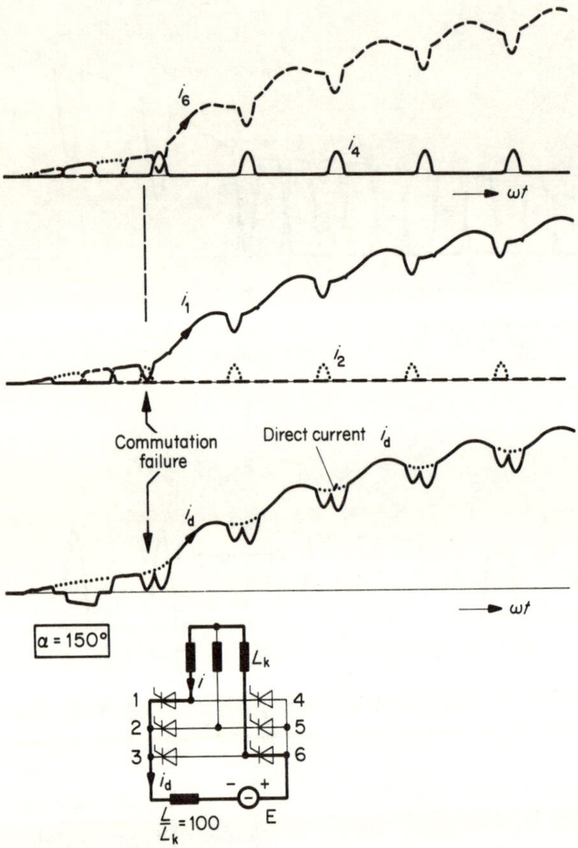

Figure 110. The inverter can lose control either through valves in a by-pass path (Figures 108 and 109) or through the three-phase system (bottom left). Example with $\alpha = 150°$ and $L/L_k = 100$. The direct current, modulated by the alternating voltage, increases continuously. Energy is supplied by the d.c. voltage source E

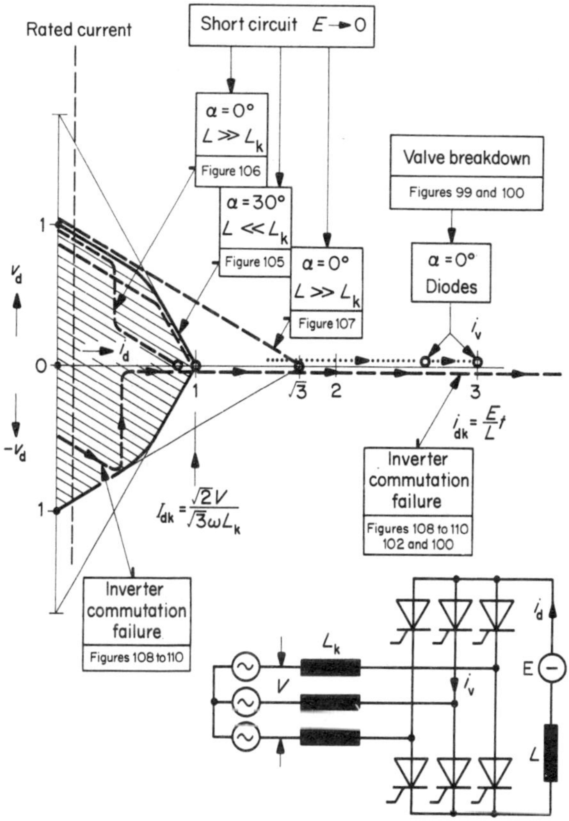

Figure 111. Summary of fault situations in the threephase bridge leading to overcurrent. Comparison may be made with the complete load diagram (Figure 63). Detailed diagrams are referred to for individual faults

11

The Three-Phase Bridge under Special Operating and Switching Conditions

In the study of the three-phase bridge, ohmic resistance and capacitance can usually be neglected, and only the inductances of the three-phase system and the d.c. circuit taken into account. In special circumstances, however, the ohmic resistances or the capacitances can affect the operation of the converter considerably. This will be illustrated by some examples.

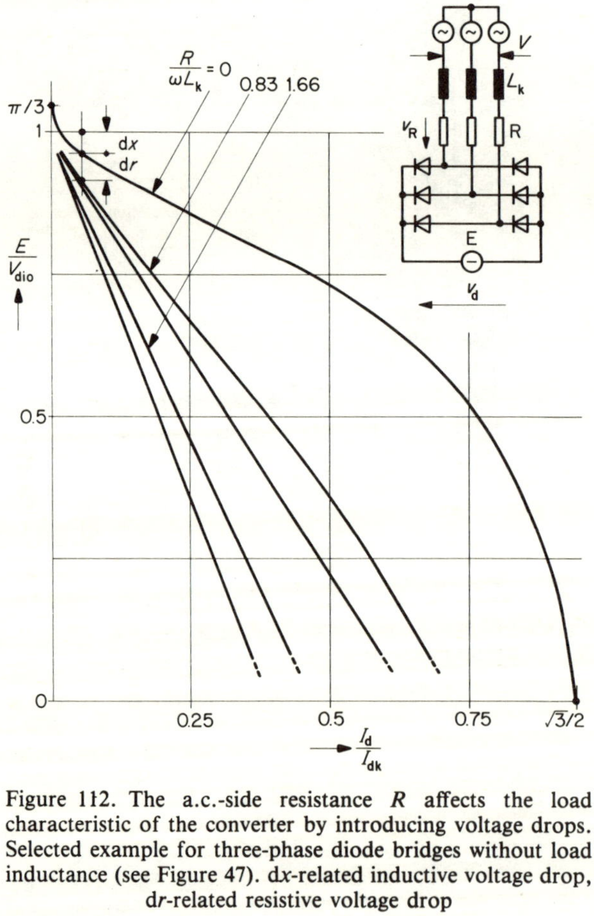

Figure 112. The a.c.-side resistance R affects the load characteristic of the converter by introducing voltage drops. Selected example for three-phase diode bridges without load inductance (see Figure 47). dx-related inductive voltage drop, dr-related resistive voltage drop

THE EFFECT OF OHMIC RESISTANCES ON LOAD CHARACTERISTICS

In high-power transformers the ohmic resistance of the windings can normally be neglected. The winding resistances of smaller transformers often need to be taken into account, however.

Equally, when a converter is used in the rotor circuit of an induction motor (the 'subsynchronous converter cascade') the ohmic resistances of the rotor winding (which represents a three-phase system of variable frequency) have a not-insignificant effect (Figure 137).

The *a.c.-side ohmic resistances* give rise to additional voltage drops, dr, which affect the load characteristics as shown in Figure 112. The direct voltage produced is lower as a result of the ohmic resistances of three-phase system.

In the inverter mode the limit of control is reached at a lower direct current than in the ideal case as a consequence of the ohmic resistances.

Resistance on the d.c. side has an effect similar to that of a back-voltage. This is illustrated by the example in Figure 113 ($\alpha = 30°$, R arbitrary), with and

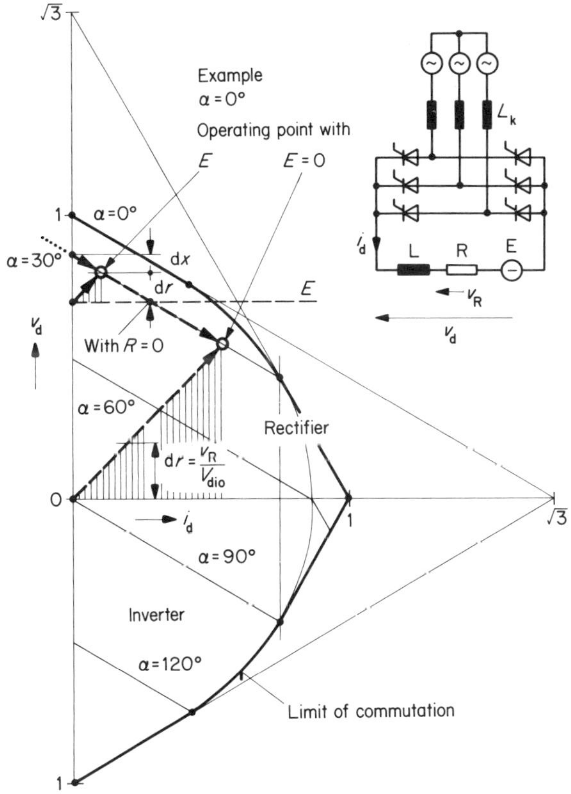

Figure 113. Ohmic resistance R in the d.c. circuit increases the load back-voltage by the (relative) value dr. Examples of the steady-state operating points associated with $\alpha = 30°$, with and without d.c. voltage source E

without a d.c. voltage source E. The ohmic voltage drop v_R increases in proportion to the load current i_d. The steady-state condition is attained when the mean value of the converter voltage V_d is equal to the sum of E and v_R.

CAPACITIVE CONVERTER LOADING

A load capacitor with capacitance C, together with the a.c.-side inductance L_k, forms an *oscillatory circuit*. If the natural frequency of this oscillatory circuit is much higher than the frequency of the alternating voltage source v, the switching-on voltage can be regarded as constant, and an equivalent circuit such as that of Figure 19 is applicable. The capacitor voltage rises to twice the applied voltage and then remains constant. The capacitor is charged by a half-sine-wave pulse of current.

If the natural frequency of the oscillatory circuit is equal to or less than the

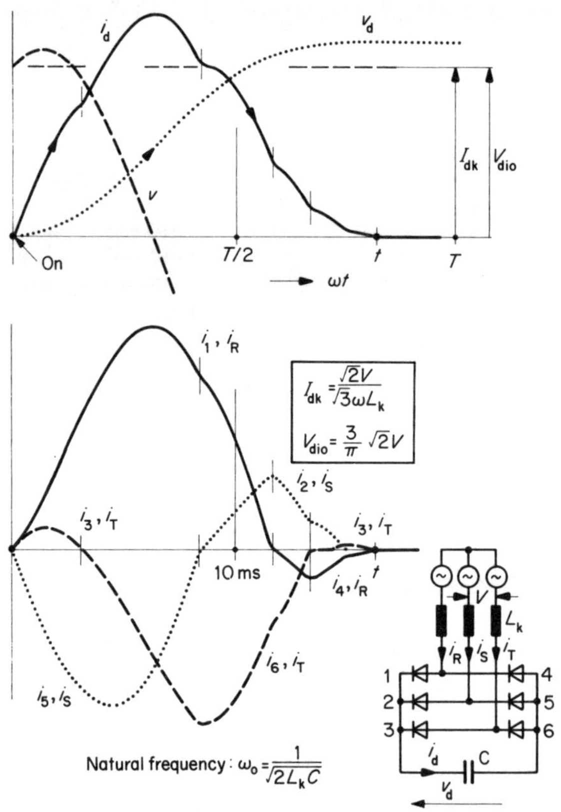

Figure 114. The switching-on process in the diode bridge with large capacitive load entails commutation. The switch-on current is higher than the steady-state short-circuit current I_{dk} (compare Figure 111). With a small load capacitance the capacitor is charged to twice the switching-on voltage (Figure 19). Example with $\omega_0/\omega = \frac{3}{4}$

supply frequency, commutation takes place in the three-phase bridge during the charging process, as illustrated in Figure 114. The capacitor voltage then increases more slowly and reaches a much lower final value than in the first case. The peak current, however, is greater than the steady-state short-circuit current I_{dk}.

THE EFFECT OF TRANSFORMER CONNECTIONS ON LOAD CHARACTERISTICS

The three-phase bridge may be supplied from a transformer whose secondary winding can be connected either in *star* or in *delta* (Figure 28). From the point of view of the converter it is the line voltage that is significant, and the waveform of the alternating current i is not dependent upon the transformer winding connection.

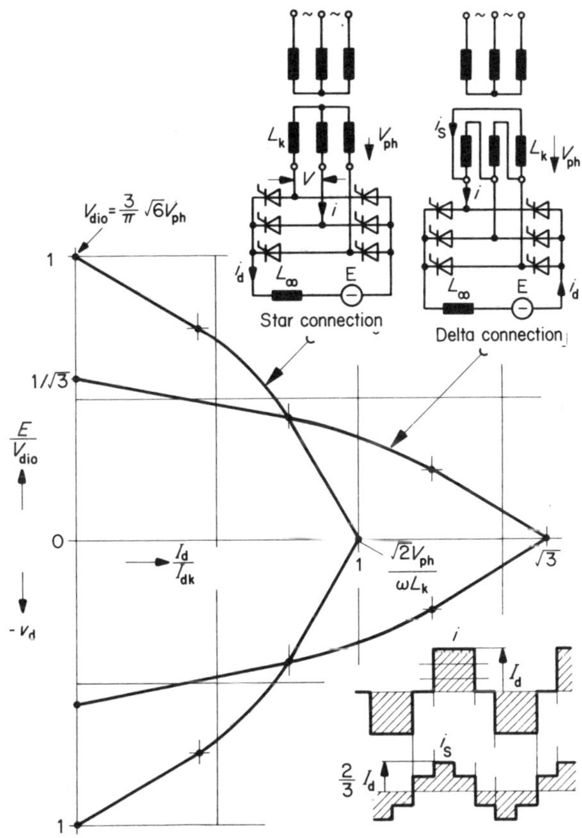

Figure 115. The three-phase bridge can equally well be fed from a transformer with secondary windings connected in star or in delta. If, however, particular windings are switched from the star to the delta connection, the load characteristics of the converter are changed

Figure 115 shows how the load characteristic of the three-phase bridge is changed when the secondary winding of the transformer is switched from the star to the delta connection. The alternating currents are shown at the bottom right of the figure in idealized form.

The harmonic content of the winding current in the star connection is the same as that in the delta connection. Similarly, the harmonic spectrum (Figure 85) is the same for both current waveforms: only the polarities of the fifth, seventh, seventeenth, nineteenth, etc., harmonics are reversed (see, further, Figures 141 and 142).

THE EFFECT OF SUPPLY SHORT-CIRCUIT CAPACITY ON LOAD CHARACTERISTICS

The load characteristics of the converter (Figures 56, 63, etc.) are determined by the total effective commutating inductance L_K of the three-phase system. The commutating inductance L_K normally consists mainly of the short-circuit inductance L_T of the transformer together with the short-circuit inductance L_s of the supply mains. The short-circuit inductance of the transformer is related to the relative short-circuit voltage e_{kt} and that of the supply mains to its short-circuit capacity Q_s. (A 'stiff' supply has a high short-circuit capacity and therefore a low short-circuit inductance, a 'weak' supply the reverse.)

The division of the short-circuit inductance between the transformer and the supply has no significance in the converter load diagrams as presented in relative terms (Figures 43, 44, 47, 56, 63, etc.). It merely shifts the nominal operating point: the load diagram is unchanged. This is shown in Figure 116 (bottom). The reference quantities are the actual values of V_{dio} and I_{dk}. The ratios of rated current I_{dN} to short-circuit current I_{dk} and of rated voltage V_{dN} to V_{dio} ($V_{dN}/V_{dio} = 1 - dx$) are modified.

Figure 116 (top) shows the load diagram, in which V_{dio} and I_{dk} are chosen as reference values for a transformer connected to an ideal supply. An increasing supply inductance L_s then reduces the short-circuit current. The relative rated current I_{dN}/I_{dk} does not change, but the direct voltage V_{dN}/V_{dio} is reduced.

For the purposes of circuit theory the rated current I_{dN} of the converter (with ideal valves) is determined by the rated current I_N of the transformer, which is related to the short-circuit voltage e_{kt} by definition. The relative short-circuit voltage is the determining factor in the location of the rated current (I_{dN}/I_{dk}) in the load diagram (Figure 56). Additional supply inductance alters the short-circuit current I_{dk}, however, and it is possible to represent the ratios differently in the load diagram, as shown in Figure 116.

The nominal working point of the converter is given by the rated voltage V_{dN} and the rated current I_{dN}. If this nominal rating is to be maintained with different short-circuit levels of the supply to which the converter is connected (through the converter transformer), the transformer secondary voltage must be increased for lower short-circuit levels. This is illustrated in Figure 117. The basic load characteristics of the three-phase bridge are shown here for two extreme cases—an ideal supply with $Q_s = \infty$ and one with the supply exactly

The above equations are valid on the following assumptions:

1. A high inductance in the d.c. circuit ($L \gg L_k$)
2. Overlap angle $\mu < 60°$
3. A transformer with rated current I_N and short-circuit voltage e_{kt} on an ideal supply (the short-circuit reactance X_T of the transformer is equal to the commutating reactance X_k)

The power rating of the transformer (apparent power) is given by

$$S_T = \sqrt{3}\, I_N V \tag{108}$$

The short-circuit level of the transformer can be defined as follows:

$$Q_T = \frac{S_T}{e_{kt}} = \frac{V^2}{X_k} \tag{109}$$

If the converter transformer is connected to a supply with a short-circuit level Q_S, the commutating reactance is increased by the amount of the supply reactance

$$X_s = \frac{V^2}{Q_S} \tag{110}$$

that is,

$$X_k = X_T + X_s = X_T\left(1 + \frac{Q_T}{Q_s}\right) \tag{111}$$

The d.c.-side voltage drop

$$\Delta V_x = \frac{3}{\pi} I_d X_k \tag{112}$$

is therefore increased (see Figures 116 and 117).

Ohmic resistances in the d.c. circuit (Figure 113) give rise to a voltage drop

$$\Delta V_R = RI_d \tag{113}$$

In relative terms:

$$dr = \frac{\Delta V_R}{V_{dio}}$$

Ohmic resistances on the a.c. side (Figure 112) also reduce the direct voltage, by a little less than

$$\Delta V_R \approx 2RI_d \tag{114}$$

12

Double-Bridge Converters

GENERAL

The fully controlled three-phase bridge with thyristors can function as a rectifier or as an inverter (Figures 63, 64, 71, 73, and 77). The *converter can, by suitable control, reverse the polarity of the direct voltage, but not the direction of the direct current.* To change the direction of current, therefore, it is necessary to use two three-phase bridges connected in inverse parallel. This circuit may be termed a *reversible converter.* The same circuit can, however, function as a frequency converter (*direct frequency changer*) if the output voltage at the d.c. terminals is varied sinusoidally by the control.

Two three-phase bridges decoupled on the d.c. side by an inductor can interconnect two three-phase systems of different frequencies and voltages, and provide for an exchange of energy in two directions. This double-bridge circuit is known as a *d.c.-link frequency changer*.

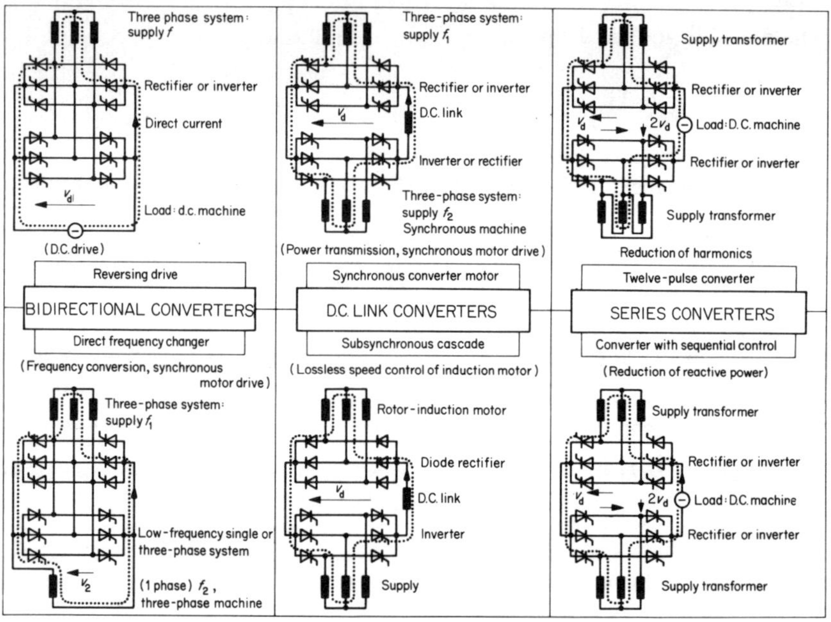

Figure 118. With two three-phase bridges, circuit and control variations lead to six converter combinations with different properties, covering the principal applications of the three-phase converter in industry and power supply engineering. In the diagram: classification of two-bridge converter circuits

The three-phase bridges may also be connected in series. The direct voltage is then doubled. At the same time, by using a star–delta transformer connection, it is possible to obtain *twelve-pulse* converter operation, with a reduction in harmonics, or to operate the converter with reduced reactive power by means of *sequential control.*

The various arrangement of converters composed of two three-phase bridges are summarized in Figure 118 (in symbolic form, see Figure 2). In the reversible converter only one bridge operates at any given time, the other being quiescent (the so-called circulating-current-free circuit). In the d.c.-link frequency changer, on the other hand, both bridges are always operating simultaneously.

A summary of the most important converter drives using two bridges is presented in Figure 119.

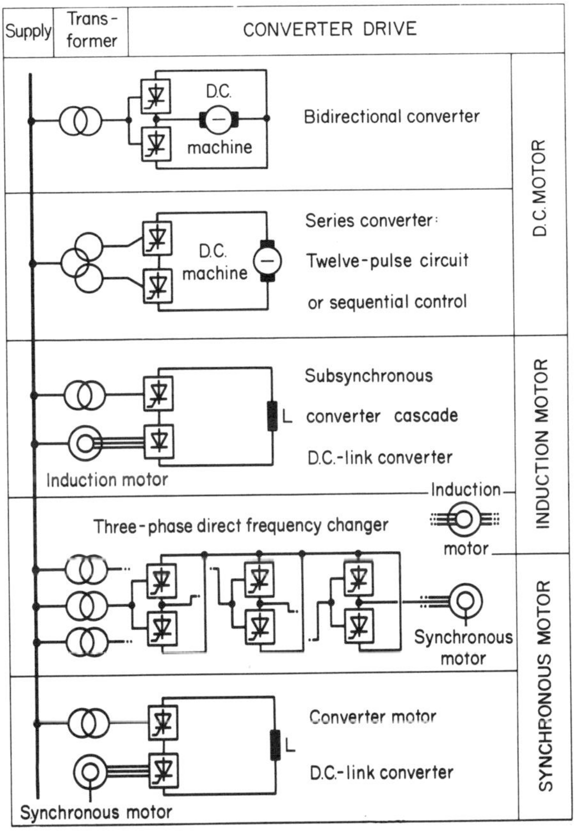

Figure 119. Converters with two three-phase bridges are used both for for d.c. drives (reversible converter, twelve-pulse converter, sequential control) and for three-phase a.c. drives (subsynchronous converter cascade, synchronous converter motor, frequency changer supplying synchronous or induction motor). Summary of basic circuits

REVERSIBLE CONVERTER DRIVES

Operating principle

The d.c. drive with a separately excited d.c. machine can be considered as generally the most important drive system. To reverse the torque of the d.c. machine, as required for the purpose of braking, or to reverse the direction of rotation, the current in either the armature or the field must be reversed. This can be achieved by two three-phase bridges *connected in inverse parallel.*

The circuit of a reversible converter consisting of two three-phase bridges is illustrated in Figure 120 (and, further, in Figure 123). According to which of the two bridges (A or B) is operating, the current i_d flows in one direction or the other in the armature circuit of the d.c. machine. The changeover logic ensures that *only one of the two bridges operates* while the other blocks.

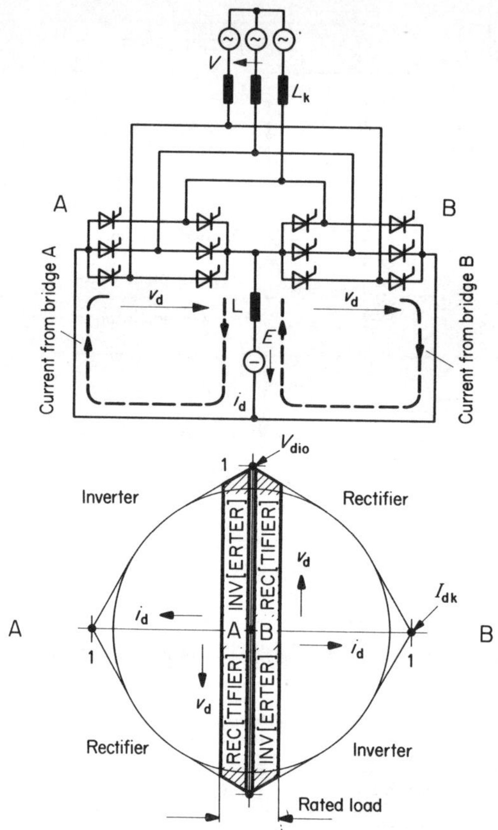

Figure 120. Two bridges connected in inverse parallel form a reversible converter. Only one bridge is conducting at any particular time ('circulating current-free' circuit). The associated d.c. machine may function as a motor or as a generator in either direction of rotation (four-quadrant operation). Top: circuit. Bottom: load characteristic

It will be seen from the load diagram (figure 120, bottom) that the reversible converter can operate in all *four quadrants* ($\pm v_d$ and $\pm i_d$). The operation of the reversing drive can be explained by reference to Figure 121.

In the period immediately following the switching-on of the drive, the direct current i_d is built up as quickly as possible in the required direction. Bridge A is therefore turned fully on ($\alpha = 0°$). The machine is as yet at standstill, and the back-voltage E in the equivalent circuit is therefore zero. When the current has reached the rated value the machine (it is assumed) begins to turn, and the load back-voltage increases in proportion to the speed n. To keep the direct current i_d at the desired constant value, the rectifier voltage v_d (the mean value must at all times match the back-voltage E. The delay angle α of bridge A is adjusted during this period from $\alpha = 90°$ ($E = 0$) to $\alpha = 0°$ (E = rated voltage). To reduce the current when the rated speed has been attained, bridge A is switched to the inverter mode. The speed n and the related back-voltage E are unchanged, however, since the torque is not reversed.

After a short *changeover* interval (it must be ensured, by measurement techniques, that the current has fallen to zero) bridge B is unblocked. The current is built up again, but in the reverse direction (bridge B operates briefly in the rectifier mode) and is then held constant. The torque, now negative, has a braking effect on the machine (an idealized linear deceleration is assumed). The voltage v_d of bridge B is adjusted accordingly: the converter operates in the *inverter mode*. The mechanical energy in the machine is fed back into the

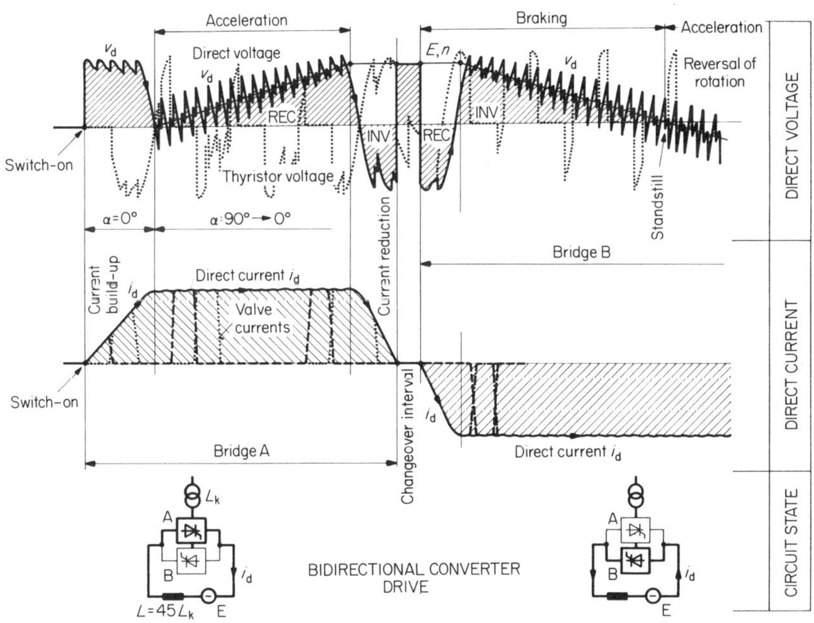

Figure 121. With the reversible converter (circuit Figure 120) the d.c. drive can operate in both directions of rotation, either as a motor or as a generator (e.g. for braking). Example of acceleration followed by braking. Current and voltage waveforms on idealized assumptions

supply mains. When the machine reaches standstill ($n = 0$, $E = 0$), the voltage v_d can be controlled further in the opposite polarity, so that the machine runs in the reverse direction. The whole process is then repeated, but with the roles of the converters and the direction of rotation reversed.

The reversible converter, in accordance with Figures 120 and 121, is described as 'circulating-current-free', and represents the simplest variant of the reversible converter circuits. If it is required to make the reversal more rapid and eliminate the changeover interval, the reversible converter can be operated in the 'circulating-current' mode. In this case both bridges are loaded simultaneously by a circulating current, which is kept within limits by an interposed d.c. inductor. In addition, each bridge is supplied from a separate transformer winding.

Fault conditions in the reversible converter

Apart from the faults that can occur in the operation of single three-phase bridge (see Figure 111), the possibility has to be taken into account in the

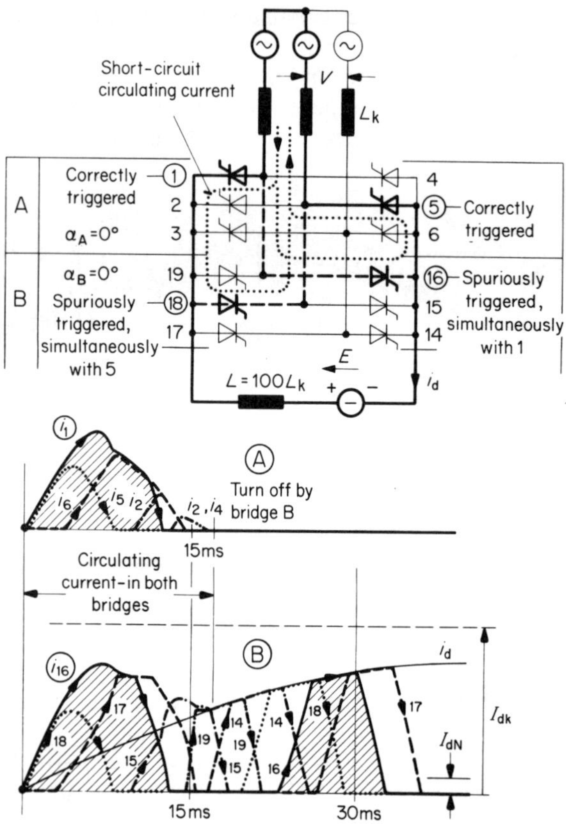

Figure 122. If both bridges of the reversible converter are (inadvertently) turned on, the result is a fault with large overcurrents ('circulating current'). Example of the fault situation with $\alpha_A = \alpha_B = 0°$. The spuriously conducting bridge B remains in the short-circuit state. Bridge A blocks

reversible converter of a *circulating short circuit,* which can occur if both bridges are inadvertently triggered simultaneously. Figure 122 shows an example. Bridge A is intentionally triggered with $\alpha = 0^\circ$; at the same time bridge B is unblocked. (Thyristor 16 is triggered simultaneously with 1, thyristor 18 with 5, etc.) In the initial period of the fault both bridges carry a current, which increases in the valves in the manner of a short circuit, since the effect is initially that of a single-phase short circuit across the transformer. Bridge B is in the state of an inverter with commutation failure. The resulting short-circuit current i_d increases continuously and forces the current in bridge A down until it blocks. The spuriously triggered bridge B then remains in the short-circuit condition.

If in an otherwise blocking converter two thyristors (e.g. 1 and 18) are triggered at the zero crossing of the voltage V, the valve current rises to an amplitude equal to $\sqrt{3}\, I_{dk}$.

The *circulating current* that flows in both bridges under the fault condition

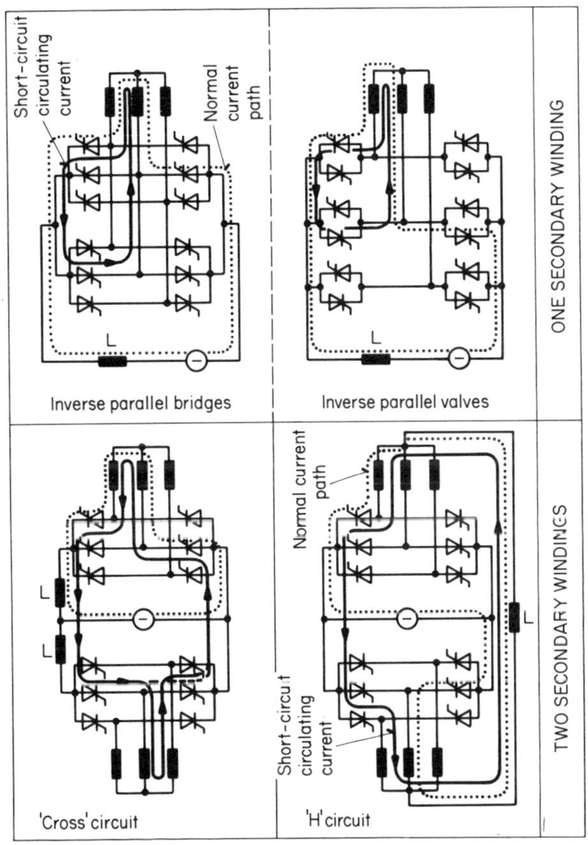

Figure 123. Apart from the simplest variants of the reversible converter with bridges or thyristors connected directly in inverse parallel (right), circuits are also used which need two transformer secondary windings (left). In the 'cross' and 'H' circuits the short-circuit circulating currents (Figure 122) are limited by the inductance L

is equal to the short-circuit current of the transformer, and is not affected by the load inductance L. If, however, the two bridges are fully separated and supplied from two secondary windings on the converter transformer, the circulating fault current can be limited by means of the so-called *'cross' circuit* (requiring two d.c. inductors) or the *'H' circuit* (with only one d.c. inductor). These variants of the reversible converter circuit are displayed in Figure 123. The 'cross' and 'H' circuits necessitate about a 20 per cent. increase in transformer rating in comparison with the inverse-parallel circuit.

DIRECT FREQUENCY CHANGER

Single-phase direct frequency changer

The circuit of the direct frequency changer is similar to that of the reversible converter (Figure 118): *two inverse-parallel-connected three-phase bridges* (or an alternative reversible converter circuit in accordance with Figure 123). The function of the converter—the facility for frequency transformation—is achieved by means of appropriate control. The unidirectional votage v_d of the two bridges is so controlled that its mean value periodically forms a sinusoidal fundamental wave of voltage v_{d1}, which produces a sinusoidal current i_2 of low frequency (relative to the supply voltage).

The operation of the direct frequency changer with single-phase output is illustrated in Figure 124. The delay angle of the bridges is so modulated that

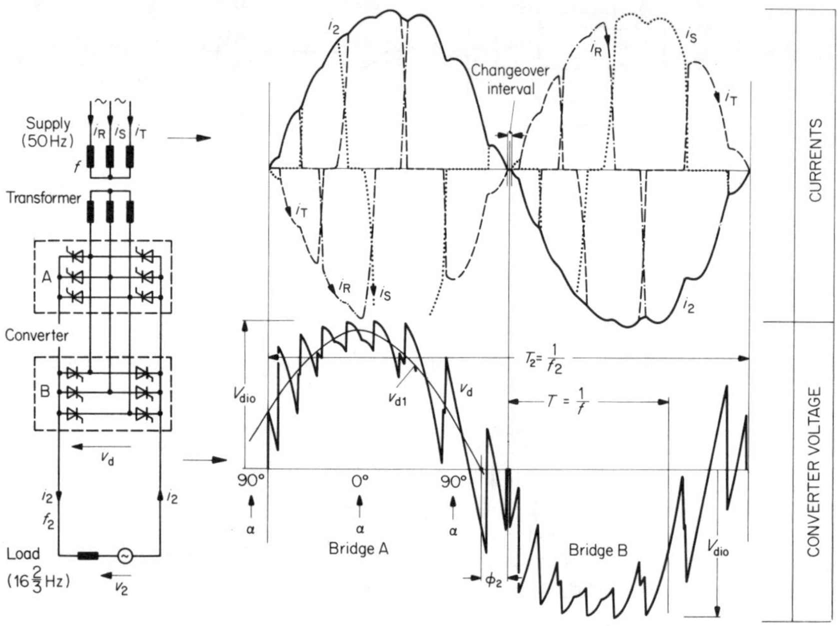

Figure 124. Two three-phase bridges in inverse parallel can be so controlled that the rectified output voltage v_d has a sinusoidal waveform. The circuit operates as a frequency changer and is referred to as a direct frequency changer. Example of a (single-phase) direct frequency changer for 50–16$\frac{2}{3}$ Hz

the unidirectional voltage v_d follows a sine wave. The current i_2 is consequently also sinusoidal. The output frequency f_2 in the example shown is one-third of the supply frequency f. Lower frequencies, from a few hertz to direct current ($f_2 = 0$), are obtainable without any modification, but frequencies higher than one-third to one-half the supply frequency cannot be achieved. *The direct frequency changer can only reduce the input frequency f.* In Figure 124 the direct frequency changer converts from 50 Hz three-phase to $16\frac{2}{3}$ Hz single-phase voltage.

The phase difference ϕ_2 between the low-frequency fundamental voltage and the low-frequency fundamental current is readily accommodated. Power transfer can therefore take place in both directions, from the three-phase system to the low-frequency single-phase system, or vice versa.

The operating range of the direct frequency changer in the characteristic area of the load diagram can be seen in the example of Figure 125. The

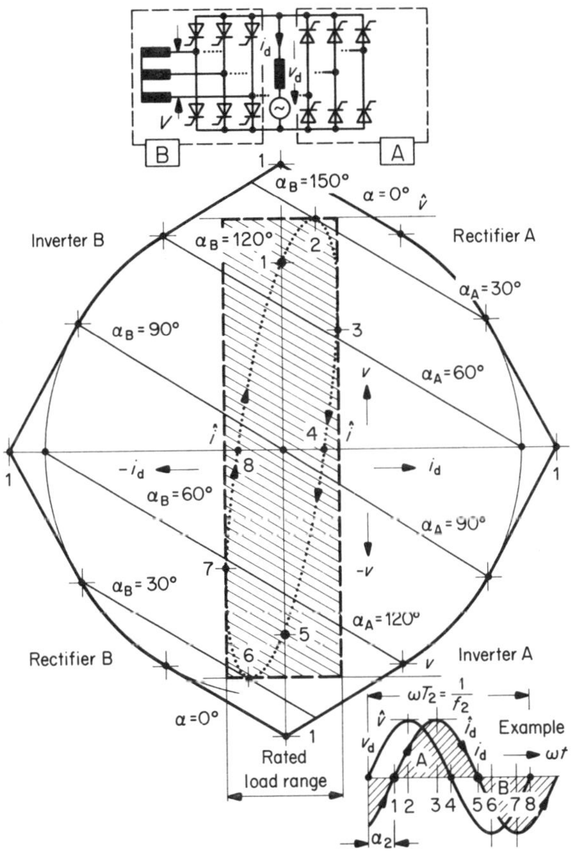

Figure 125. The direct frequency changer operates in all four quadrants of the load diagram in the course of one low-frequency cycle (exception: $\phi_2 = 0°$). Example with $\phi_2 = 60°$. Both bridges function periodically at different times in both the rectifier and the inverter mode

changeover from one three-phase bridge to the other takes place in all instances at a current zero (instants 1 and 5).

The current in the three-phase system of the direct frequency changer (in the transformer and the supply) exhibits basically the characteristics of the alternating current in the three-phase bridge (Figures 75, 79, 83, etc.), but is low-frequency-modulated in amplitude, as can be seen in Figure 124. The result of this is that the active and non-active power in the three-phase system are also pulsed at low frequency (Figure 129) and, further, that the current in the three-phase system contains a much wider spectrum of harmonics than that of the normal converter (see also Figures 126 to 128).

The amplitude spectrum of the steady-state alternating current of the converter contains only harmonics of orders 5, 7, 11, 13, etc., regardless of whether the load is resistive (Figure 90), inductive (Figure 89), or mixed

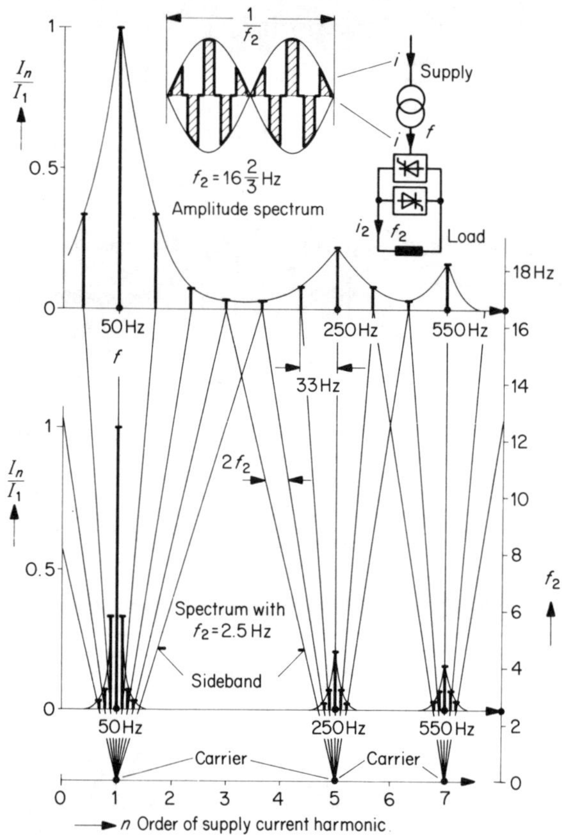

Figure 126. As well as the harmonics of the characteristic orders $n = 5, 7, 11, 13, \ldots$, the supply current of the single-phase direct frequency changer contains sidebands with frequencies at intervals of $2f_2$. Example of amplitude spectrum for $f_2 = 2.5$ Hz and $16\frac{2}{3}$ Hz. Operation in the sinusoidal mode with maximum output: $\phi_2 = 0°$

(Figure 91). Harmonics of other than the characteristic frequencies arise only in transient states of the three-phase bridge (Figures 96 and 97).

The supply current of the direct frequency changer includes, besides the characteristic harmonics (the carriers) in accordance with the basic law $n = k6 \pm 1$, sidebands of each carrier frequency, whose frequencies depend upon the low-frequency modulation. Thus the fifth harmonic is split into the following spectrum: 250 Hz, $250 \pm 2f_2$, $250 \pm 4f_2$, $250 \pm 6f_2$, etc. The same adjacent frequencies $\pm 2f_2$, $\pm 4f_2$, $\pm 6f_2$, etc., are produced around each carrier (including the fundamental component).

The splitting of the characteristic harmonics and the fundamental component into sidebands is illustrated in Figure 126. The lower the output frequency f_2, the narrower are the groups of sidebands. With d.c. output ($f_2 = 0$) only the carriers ($n = 1, 5, 7, 11, 13, \ldots$) remain.

Three-phase direct frequency changer

Three direct frequency changers (in accordance with Figure 124) can be used to form a *low-frequency three-phase system,* provided that the individual output voltages are mutually displaced by 120° and that the three converter phases are appropriately combined.

The three-phase direct frequency changer is illustrated in Figure 127. In the supply current i_s, consisting of the combined transformer currents of the three

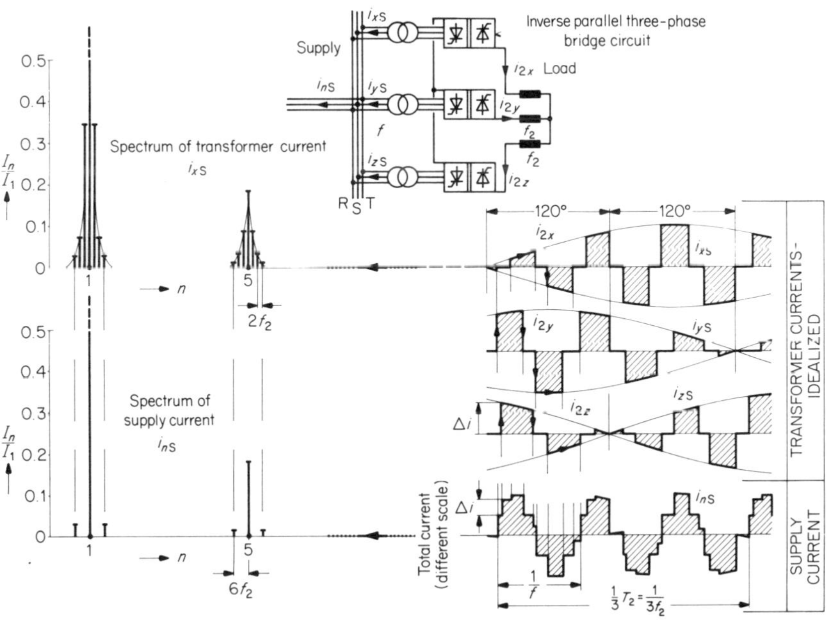

Figure 127. A low-frequency three-phase system can be constructed with three single-phase direct frequency changers. The supply current i_s in the three-phase direct frequency changer is composed of the three transformer currents mutually displaced by 120°. Left; comparison of the current spectra

individual direct frequency changer phases, the low-frequency modulation is hardly discernible. Two-thirds of the sidebands present in the transformer currents are omitted from the harmonic spectrum of the supply current. The sidebands at $\pm 2f_2$, $\pm 4f_2$, $\pm 8f_2$, etc., disappear, and only those at $\pm 6f_2$, $\pm 12f_2$, etc., remain (Figure 127, left).

The frequencies of the sidebands depend upon the output frequency f_2. The amplitudes of the harmonics, on the other hand, are determined by the depth of modulation and by the phase angle ϕ_2.

So long as the amplitude of the fundamental voltage v_{2m} is less than the ideal direct voltage V_{dio}, the converter may be described as a 'sinusoidal converter'. If the fundamental amplitude of the output voltage exceeds the value V_{dio}, the

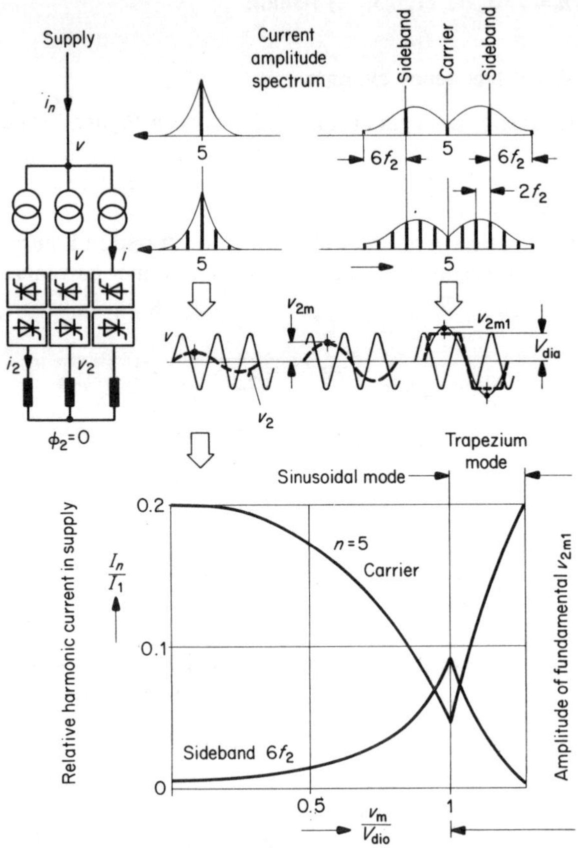

Figure 128. The amplitudes of the supply current harmonics in the three-phase direct frequency changer depend upon the depth of modulation (v_{2m}/V_{dio}) and the phase angle ϕ_2. As an example, the variation with modulation depth of the fifth harmonics and their first sidebands (with $\phi_2 = 0\,°$). v_m = amplitude of the fundamental component v_{d1} according to Figure 124. In the sinusoidal mode $v_m = v_{2m}$; in the trapezoidal mode $v_m = v_{2m1}$

output voltage assumes a trapezoidal waveform, and the arrangement may then be termed a 'trapezoidal converter'.

The variation in amplitude of the fifth harmonic carrier and its first sidebands over the whole control range of the three-phase direct frequency changer (with $\phi_2 = 0^\circ$) is illustrated in Figure 128. It will be observed that the sidebands can reach a greater amplitude than the carrier.

The total distortion of the supply current (the harmonic content) is nevertheless smaller in the three-phase direct frequency changer than in the simple bridge circuit.

The harmonic spectrum of the supply current in the direct frequency changer is given by Fourier analysis. The spectrum including the sidebands (according to Figures 126 to 128) is, however, obtained only when the analysis

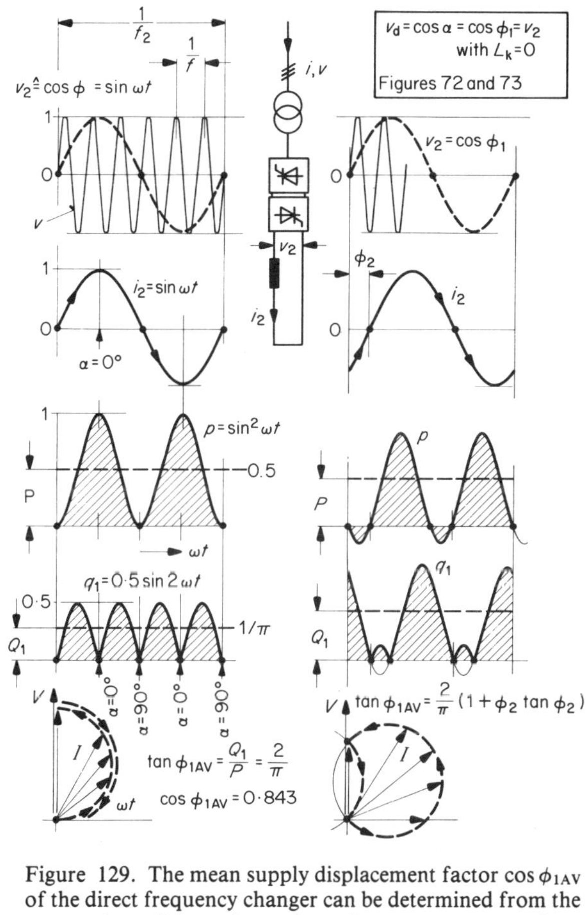

Figure 129. The mean supply displacement factor $\cos \phi_{1AV}$ of the direct frequency changer can be determined from the mean values of the active and reactive powers over one low-frequency cycle. Example with $v_{2m}/v_{dio} = 1$ and $\phi_2 = 0°$ and 45°. Commutation neglected. The dependence of the supply displacement factor upon $\cos \phi_2$ and the modulation depth is illustrated in Figure 130

is carried out over a period corresponding to the low-frequency cycle ($T_2 = 1/f_2$). If, for example, the supply current i_s is analysed over one supply cycle $T = 1/f$, the spectrum will appear to be quite different, and will moreover vary from cycle to cycle, since the current waveform does not stay constant.

When a direct frequency changer is used to supply the rotor of an induction motor (Figure 119) harmonic currents are induced in the stator from the rotor, leading to an increased harmonic loading of the supply.

The three-phase bridges of the direct frequency changer operate continuously ('sinusoidal') or at least in part ('trapezoidal') at less than maximum output. The supply current i_s therefore lags the supply voltage v_s by ϕ_1, even if there is no phase shift on the low-frequency side ($\phi_2 = 0°$). Since the delay angle is varying continuously, the phase shift ϕ_1 is different in each supply period. Figure 129 illustrates the power relationships in the ideal single-phase

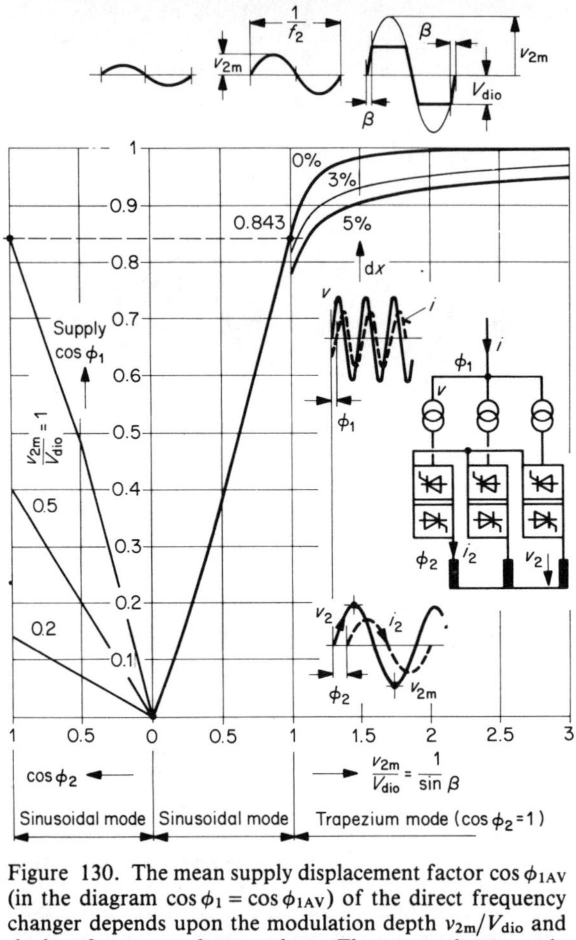

Figure 130. The mean supply displacement factor $\cos\phi_{1AV}$ (in the diagram $\cos\phi_1 = \cos\phi_{1AV}$) of the direct frequency changer depends upon the modulation depth v_{2m}/V_{dio} and the low-frequency phase angle ϕ_2. The curves shown apply to both the three-phase and the single-phase direct frequency changers. For the determination of $\cos\phi_{1AV}$ see Figure 129

direct frequency changer (no commutation, ideal control) for two conditions ($\phi_2 = 0°$ and $45°$). The mean displacement factor cos ϕ_{1AV} can be obtained from the mean values of the active and reactive powers calculated over one low-frequency cycle $1/f_2$. The dependence of the supply displacement factor cos ϕ_{1AV} upon the depth of modulation (v_{2m}/V_{dio}) and the low-frequency phase angle ϕ_2 (or cos ϕ_2) is illustrated in Figure 130. The figure applies equally to the single-phase and the three-phase direct frequency changer. The fundamental frequency reactive power Q_1 (relative to the active power) is thus the same for the single-phase and three-phase direct frequency changers. On the other hand, the total non-active power is relatively less in the three-phase than in the single-phase direct frequency changer, since the harmonics in the supply current are much smaller. The power factor λ of the three-phase direct frequency changer is higher than that of the single phase.

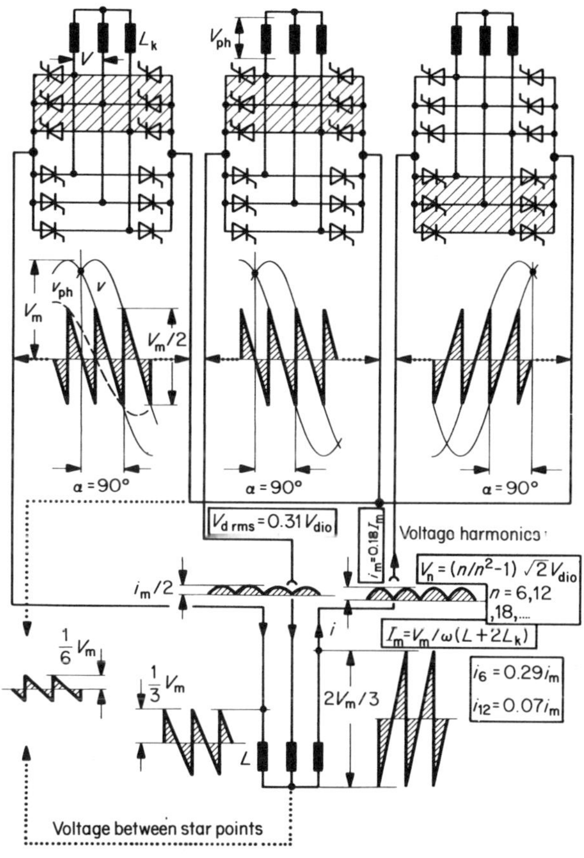

Figure 131. In the three-phase direct frequency changer with inductive load the limit of discontinuity is reached with a delay angle $\alpha = 90°$ (in all phases), as in the simple three-phase bridge (Figure 67). The maximum current ripple (Figure 69) is, however, 30 per cent greater. Diagram: voltage and current and their harmonic analyses

The direct frequency changer may operate in similar modes to those of a simple three-phase bridge (Figure 71): the load current may be discontinuous or continuous. With a purely inductive load, the limit of discontinuity is reached at $\alpha = 90^\circ$ (figure 67). The current and voltage relationships in the three-phase direct frequency changer with $\alpha = 90^\circ$ are shown in Figure 131. In comparison with the simple three-phase bridge the ripple in the load current is 33 per cent. greater.

In the trapezoidal converter the voltage waveform can be regarded as optimal when $\beta = 30^\circ$ (Figure 132). This condition gives the smallest harmonic content in the line voltage. This is especially significant when the load is the stator winding of a synchronous motor. Steeper trapezoidal waveforms can lead to overvoltages.

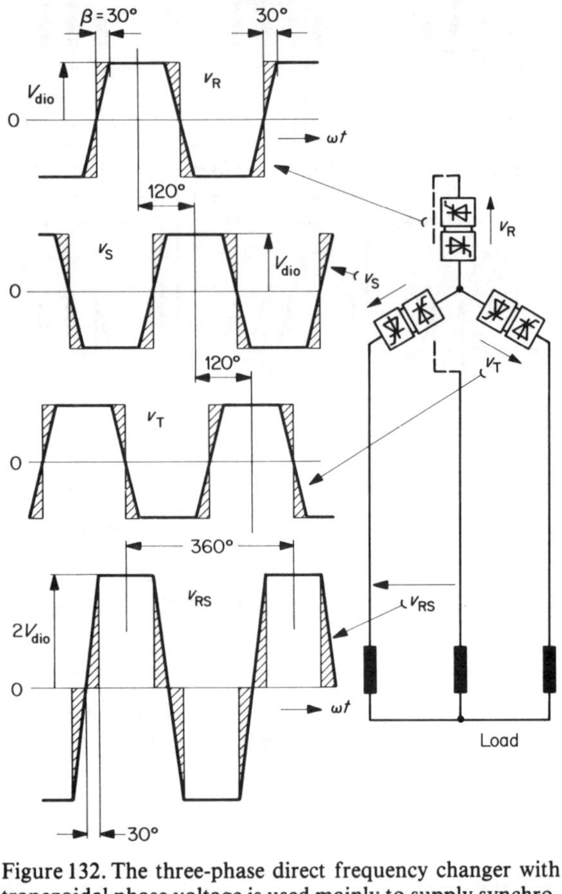

Figure 132. The three-phase direct frequency changer with trapezoidal phase voltage is used mainly to supply synchronous machines (stator winding). The trapeze angle β is made not less than 30 °, since otherwise the harmonic content of the line voltage becomes excessive, which can result in overvoltages

Fault conditions in direct frequency changers

The direct frequency changer usually consists of three reversible converters with inverse-parallel-connected three-phase bridges, and therefore exhibits basically the same faults as the reversible converter and the three-phase bridge. The following fault situations may therefore arise (see Figures 111 and 122):

1. *Breakdown* of one or more valves
2. *Terminal short circuit* on the converter
3. *Loss of control* of the converter in the inverter mode
4. *Circulating current build-up*

The cause of the fault and its termination depend upon the nature of the load. The reversible converter mostly supplies d.c. machines, whereas direct frequency changers are used in drives with three-phase a.c. machines—synchronous (Figure 119) or asynchronous.

MATHEMATICAL APPENDIX XVII

The active power in the three-phase direct frequency changer (Figure 130) on the low-frequency side

$$P = 3V_2I_2\cos\phi_2 \tag{115}$$

is equal to the active power on the supply side

$$P = \sqrt{3}\,VI_{s1}\cos\phi_{1Av} \tag{116}$$

where V = supply line voltage
V_2 = low-frequency phase voltage (fundamental)
I_{s1} = fundamental component of supply current
ϕ_2 = low-frequency-side phase angle
ϕ_{1AV} = mean supply-side phase angle

Assumptions: transformer ratio = 1:1, short-circuit inductances neglected, output frequency much lower than the supply frequency, load current I_2 sinusoidal

In ideal trapezoidal operation the output voltage is of rectilinear waveform, and the fundamental component therefore has an r.m.s. value

$$V_2 = \frac{12}{\pi^2}V \tag{117}$$

In ideal sinusoidal operation (the peak value of the output voltage v_{2m} equal to the ideal rectifier voltage V_{dio}) the low-frequency fundamental component is given by

$$V_2 = \frac{3}{\pi}V \tag{118}$$

The r.m.s. transformer current (Figure 127) is

$$I = \sqrt{\frac{2}{3}}\,I_2 \tag{119}$$

The displacement factor $\cos\phi_{1AV}$ of the three-phase direct frequency changer is equal to that of the single-phase direct frequency changer (Figure 129), and for the ideal sinusoidal mode (Figure 130) is

$$\cos\phi_{1AV} = 0.843 \tag{120}$$

From equations (115) and (116) (substituting from equation (118)), the fundamental component of the supply current (with ideal sinusoidal operation) is given by

$$I_{s1} = \frac{3\sqrt{3}}{\pi}\frac{\cos\phi_2}{\cos\phi_{1AV}} I_2 \tag{121}$$

Therefore, for $\phi_2 = 0°$,

$$I_{s1} = 1.96 I_2$$

Similarly, in the ideal trapezoidal converter (substituting from equation 117),

$$I_{s1} = 2.1 I_2$$

The ideal d.c. power from the converter according to equation (69) is defined as follows:

$$P_{dio} = V_{dio} I_d \tag{122}$$

Here the ideal direct voltage

$$V_{dio} = \frac{3}{\pi}\sqrt{2}\, V$$

and the converter direct current

$$I_d = \sqrt{\frac{3}{2}}\, I$$

are derived from the transformer parameters V and I.

Since the three-phase direct frequency changer comprises three units (three pairs of three-phase bridges), the total ideal d.c. power at maximum output, substituting from equations (118) and (119), is

$$P_{dio} = 3\sqrt{2}\, V_2 I_2 = \sqrt{2}\, P_0 \tag{123}$$

where P_0 = ideal maximum active power
$I_2 = I_d$: assumed (without any regard to the thermal time constants of the thyristors)

In ideal trapezoidal operation

$$V_{dio} = \frac{3}{\pi}\sqrt{2}\, V = \frac{3}{\pi}\sqrt{2}\,\frac{\pi^2}{12} = \frac{\sqrt{2}\,\pi}{4} V_2$$

and therefore

$$P_{dio} = 3\,\frac{\sqrt{2}\,\pi}{4} V_2 I_2 = 1.11\, P_0 \tag{124}$$

The aggregate apparent power in the transformers (in both cases) is

$$S_t = 3\sqrt{3}\,VI = \frac{\pi}{3}P_{dio} = \frac{\pi}{3}\sqrt{2}\,P_0 \tag{125}$$

The fundamental component of the trapezoidal wave voltage of one phase is given by

$$V_1 = \frac{2\sqrt{2}\sin\beta}{\pi\beta}\,V_{dio} \tag{126}$$

For β see Figures 130 and 132, (top).

For the ideal rectilinear wave (commutation neglected),

$$V_1 = \frac{2\sqrt{2}}{\pi}\,V_{dio}$$

INTERMEDIATE-LINK FREQUENCY CHANGER

Operation of the Intermediate-Link Frequency Changer

The intermediate-link frequency changer consists of two three-phase bridges decoupled by an inductor in the d.c. circuit. In the subsynchronous converter cascade only one bridge is controlled, whereas in the converter motor, and in

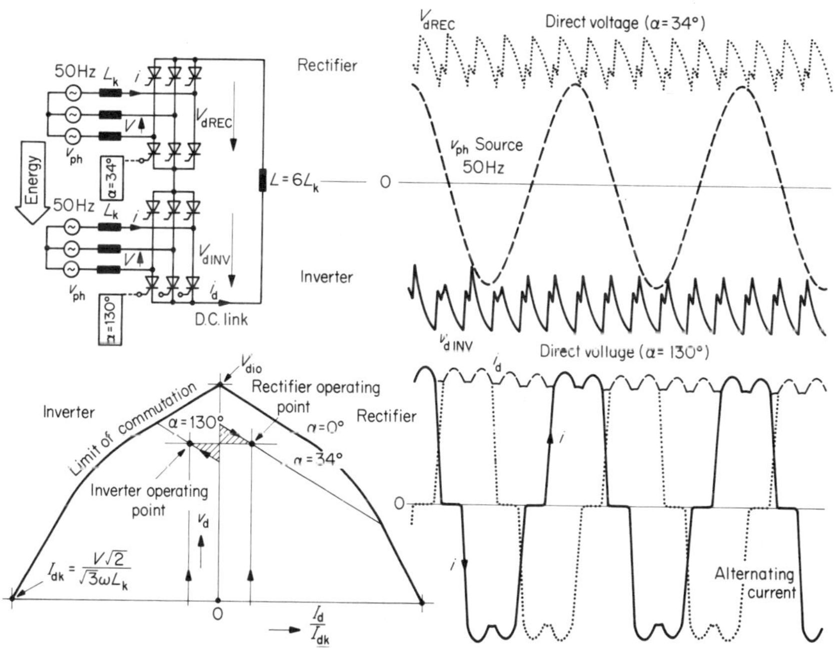

Figure 133. Two separately fed three-phase bridges, decoupled on the d.c. side by an inductance, form a d.c.-link converter (Figure 118). One bridge functions as a rectifier, the other as an inverter. Example of an operating condition with two identical three-phase systems. Characteristics, current, and voltage waveforms

power transmission, both are controlled. Figures 118 and 119 show the place of the intermediate-link frequency changer in the classification of double-bridge converters. The circuit is also known as the *d.c.-link converter,* because a direct current, smoothed by the inductor, flows between the two bridges.

The operation of the d.c.-link frequency changer, unlike that of the reversible converter (Figure 120) or the direct frequency changer (Figure 124), in which the bridges operate alternately one at a time, is based upon *simultaneous current and voltage loading of both bridges.* One bridge functions as a rectifier (Figures 43, 56 to 59), the other as an inverter (Figures 60 to 64).

Figure 133 illustrates the operation of the fully controlled d.c.-link converter in the case where the two three-phase systems are similar (the same frequency, voltage, and short-circuit inductances). The operating point of the rectifier is taken to be with $\alpha = 34^\circ$. The delay angle of the inverter then has to set at 130°

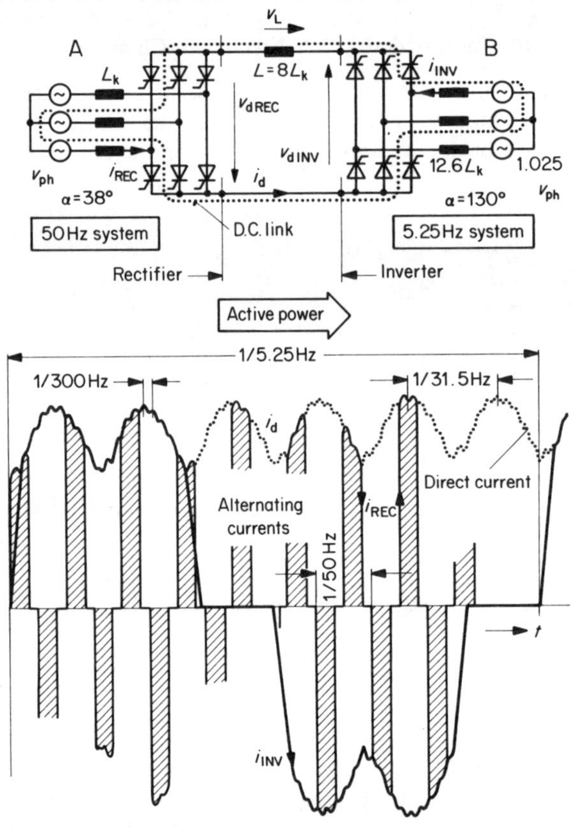

Figure 134. The d.c.-link converter normally couples two different three-phase systems. As an example, current waveforms in systems with significantly different frequencies (50 Hz/5.25 Hz). Steady-state condition. Direct and alternating currents in the two systems. Compare with Figure 135

to obtain the required current at the given voltage (see the load diagram, bottom left). The direct voltages, the alternating currents, and the direct current are shown at the right of the figure.

The two three-phase systems connected by the d.c.-ink converter can be of quite different frequencies and short-circuit inductances, and also of unequal voltages.

Figure 134 shows the currents in the d.c.-link converter for a case in which the two three-phase systems have frequencies of 50 Hz (rectifier) and 5.25 Hz (inverter). The alternating currents i_{INV} of the inverter and i_{REC} of the rectifier, and the direct current i_d, are modulated at six times the frequencies of both a.c. systems.

If the frequencies of the two a.c. systems are only slightly different, beats are noticeable in the direct current. As an example, Figure 135 shows the direct current waveform for frequencies of 50 Hz and 52.5 Hz.

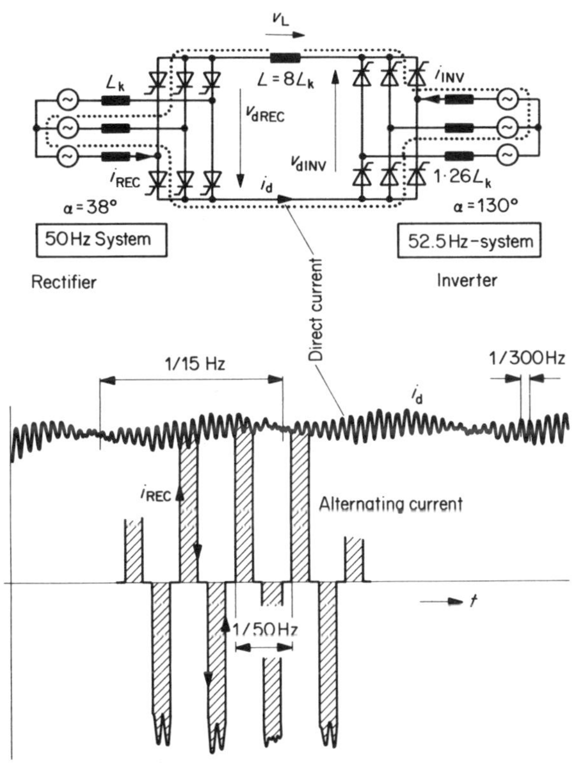

Figure 135. In a d.c.-link converter coupling two three-phase systems with only slightly differing frequencies, beats appear in the direct current. Example with frequencies of 50 Hz and 52.5 Hz. Alternating and direct currents in the rectifier. Compare with Figure 134. Uncharacteristic as well as characteristic harmonics appear in the alternating current, at frequencies which depend upon the combined effects of the two systems

Power balance in the d.c.-link converter

In the frequency changer with two controlled three-phase bridges and an inductive d.c. link (Figure 134), *the active power can be translated between the two three-phase systems in either direction.* The direction of power flow is determined by the manner in which the bridges are controlled. The rectifier side functions as an energy source, the inverter side as a load (see Figure 64).

The power transfer takes place between the a.c. voltage sources of the two systems. Only the fundamental components of the currents contribute to the power. The active power in the two three-phase systems is always the same, but the non-active powers can differ considerably. Figure 136 shows as an example the power relationships in a d.c.-link converter in which the load diagrams of the two bridges, A and B, differ in such a way (left) that the fundamental frequency reactive power Q_{1B} in the inverter is greater than that in the rectifier (Q_{1A}). The corresponding phasor diagram (for the fundamental components; see Figures 79 and 80) is shown at the right of the figure.

D.c.-link converter for three-phase drives

In drive applications the d.c.–link converter in combination with an induction motor is referred to as a *subsynchronous converter cascade* and in combination with a synchronous machine as a *converter motor* (see Figure 119).

These two applications of the d.c.-link converter are compared in Figure 137. In the subsynchronous converter cascade the rectifier (uncontrolled) is

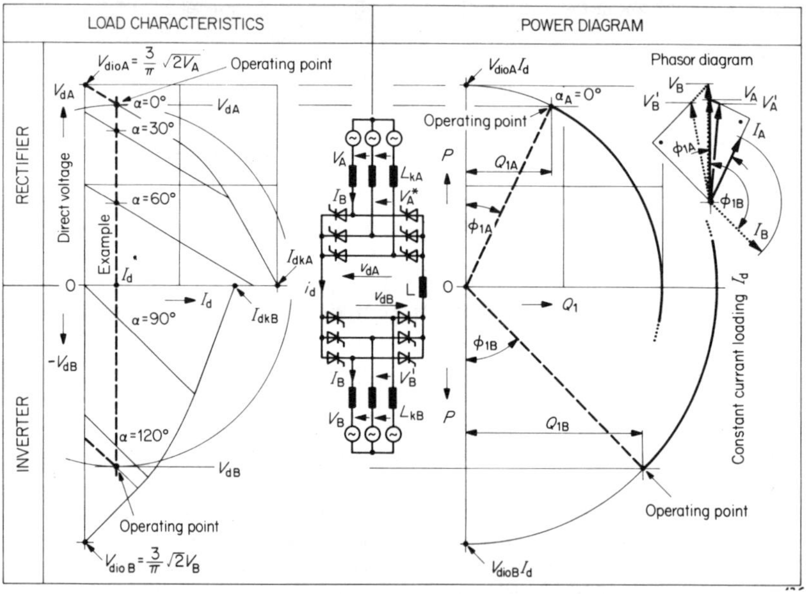

Figure 136. In the d.c.-link converter the active power is the same in both three-phase systems and identical (neglecting losses) with the active power in the d.c. link. With dissimilar three-phase systems, however, the non-active power demands are unequal. Compare with Figure 77

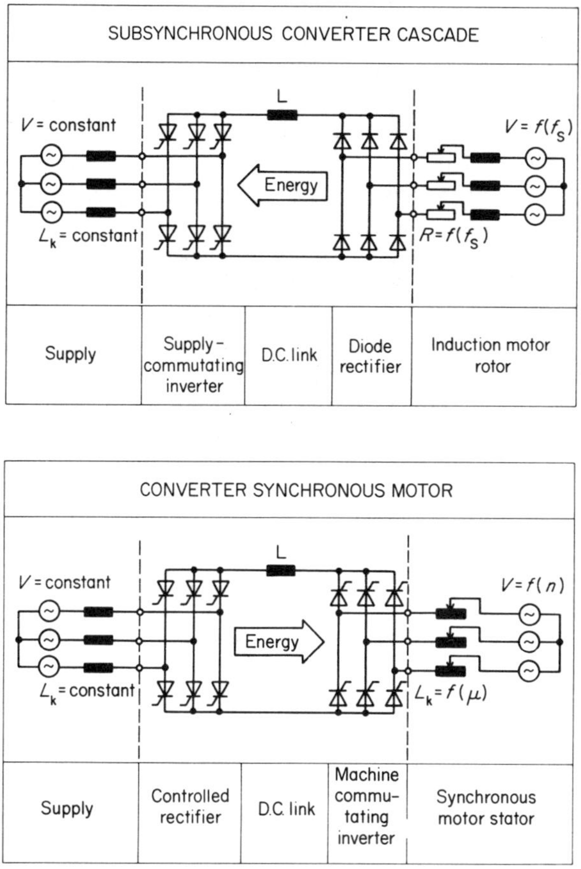

Figure 137. In drive engineering the d.c.-link converter is used mainly in the subsynchronous converter cascade or in the synchronous converter motor. Comparison of the two circuits. In the equivalent circuit of the subsynchronous cascade, ohmic resistance (Figure 112) must be taken into account, and in the converter motor, variable inductances in the commutation circuit

connected to the rotor winding of the induction motor and the inverter is connected to the supply mains. In the converter motor, on the other hand, the rectifier is connected to the supply and the inverter to the machine. The main parameters of the supply system (voltage, frequency, and short-circuit inductance) can be assumed to be constant, so that the load diagram of Figures 56, 63, etc., can be applied to the supply-side converter without any problem. For the machine-side three-phase system, however, the specific properties of the machine have to be taken into account. Thus the ohmic resistance of the rotor of the asynchronous machine cannot here be assumed negligible (for the effect on the load characteristics of the rectifier see Figure 112) and the voltage is proportional to frequency. With the machine at standstill the rotor frequency is equal to the stator (supply) frequency and the voltage (according to the transformation ratio) is at a maximum. At synchronous speed the rotor frequency and the rotor voltage are both zero.

Similarly, in the case of the synchronous machine the stator voltage is proportional to the frequency, but here the stator voltage and frequency are zero when the machine is at standstill and attain the nominal values at synchronous speed. The commutating inductance L_K, which corresponds approximately to the subtransient short-circuit inductance of the machine, is in this case not constant, but time-dependent.

The inverter of the converter motor is controlled by the machine and commutation is effected by the internal stator voltage. At standstill, however, the stator winding represents a pure inductance, without voltage sources, and inverter operation is therefore not possible. The d.c.-link converter will function only when the motor is rotating and producing a three-phase system of voltages. Acceleration of the synchronous machine from standstill is therefore contrived by means of *pulse operation*. Commutation in the inverter is simulated by the rise and fall of current in the rectifier. The direct current is

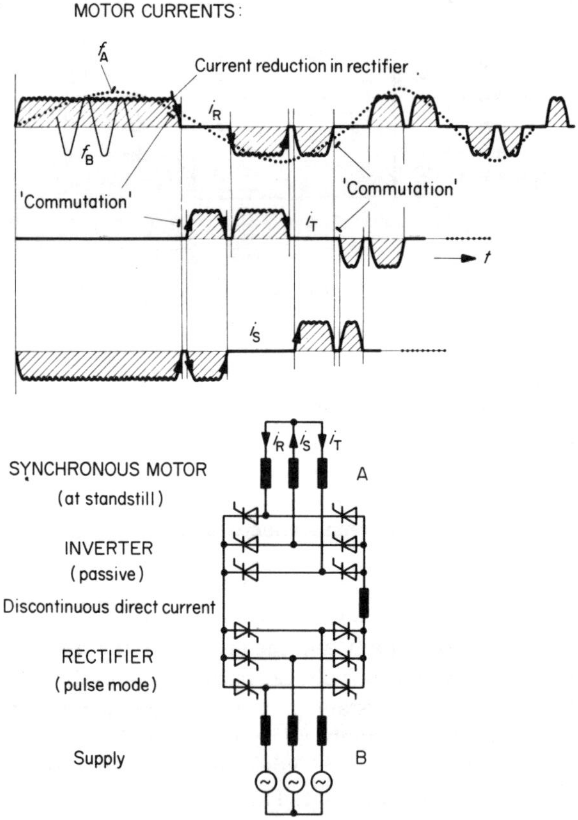

Figure 138. The problem of inverter commutation in the converter motor at standstill can be overcome, and the machine set in motion, by pulse operation of the rectifier. The synchronous machine at standstill represents only pure inductances, with no voltage sources

discontinuous, as in normal discontinuous current operation (Figure 67), but with a different time scale. Current transfer takes place in the inverter without overlap. The inverter operates in a passive mode. Acceleration in the pulsed mode is illustrated in Figure 138.

In the converter motor harmonic currents from the motor stator are coupled into the supply only through the d.c. link. In the subsynchronous converter cascade the rotor-side harmonic currents are also transformed into the stator through the air-gap, and thence into the supply. The cascade system therefore loads the supply with speed-related uncharacteristic harmonics to a much greater extent than the converter motor. In addition, beats are evident in the stator current.

Fault conditions in the d.c.-link converter

The blocking of a valve in the inverter section of the d.c.-link converter (through gate-pulse failure or a fuse interruption) can result in loss of inverter

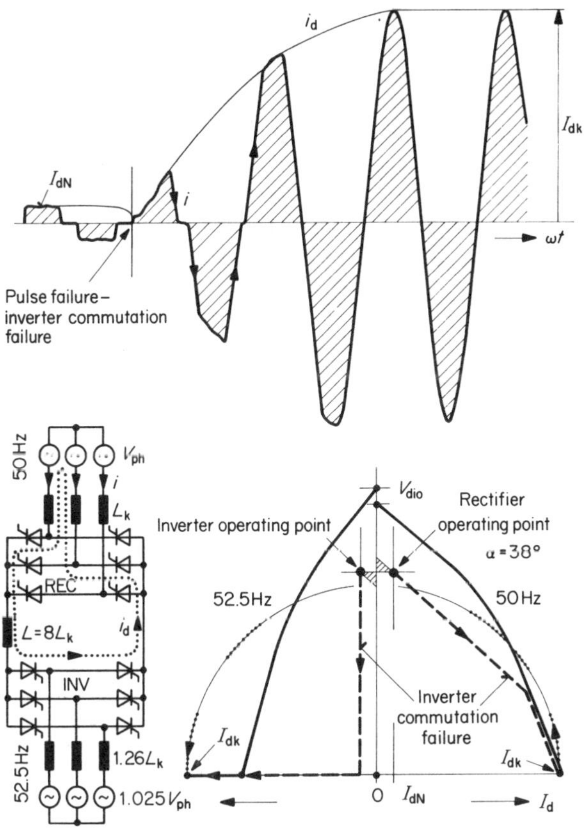

Figure 139. Loss of triggering pulses (or disconnection of a valve by a fuse) in the inverter section of the d.c.–link converter can lead to loss of inverter control. The maximum short-circuit current is then determined by the short-circuit current of the rectifier. Example: constant delay angle

control (see also Figure 102). Simultaneously the rectifier is placed in a short-circuit condition and feeds short-circuit current into the failed inverter. The maximum current loading is that represented by a rectifier short circuit. An example of this is illustrated (with a constant delay angle throughout the fault and with the parameters of Figure 135) in Figure 139.

Loss of inverter control due to alternating voltage dips, which may well be the most likely cause of inverter failure in d.c. drive (Figures 108 to 110), is less of a problem in the converter motor.

Other fault conditions in the d.c.-link converter arise in the same way as in the simple three-phase bridge (Figures 99 to 111).

Comparison of the d.c-link converter with the direct frequency changer

Both the d.c.-link converter and the direct frequency changer can be used for power translation between two different three-phase systems (supply systems or three-phase machines) (Figure 140).

In the case of the d.c.-link converter the two three-phase systems are decoupled by the d.c. link and are therefore completely independent. There is no restriction on the *frequencies* of the systems. In the direct frequency changer, on the other hand, the frequency ratio must never be less than about 3:1. (Considered from the a.c. side of the three-phase bridges, the output frequency at the d.c. terminals can only be lower than the supply frequency.)

The direct frequency changer generates the low-frequency three-phase supply of itself, functions as a low-frequency voltage source, and will supply a sinusoidal current. Only *one supply system* is necessary for its operation. The normal mode of operation of the d.c.-link converter requires *two voltage systems.* The inverter draws reactive power in commutating and its alternating current waveform is basically rectilinear. A three-phase system on the inverter side of the d.c.-link converter can obviously be started up only by pulsed operation of the rectifier (Figure 138).

The two three-phase systems of the d.c.-link converter are loaded through the commutation processes in the two three-phase bridges, and the alternating currents contain the characteristic *harmonics* of the six-pulse converter. The low-frequency current of the direct frequency changer, on the other hand, can be produced almost without harmonics and sinusoidal, but uncharacteristic harmonics and sidebands are present in the current drawn by the high-frequency supply system. The direct frequency changer, unlike the d.c.-link converter, requires no *reactive power* on the low-frequency side, but the high-frequency system is loaded intermittently by higher reactive power than in the case of the d.c.-link converter.

The three-phase direct frequency changer comprises six three-phase bridges, the d.c.-link converter only two. The required total *converter power rating* (relative to the same active power) is nevertheless greater in the d.c.-link converter than in the direct frequency changer (given ideal control). The two bridges of the d.c.-link converter carry a current continuously in the region of the peak alternating current. The correspondence of voltages is in terms of the line voltages of the three-phase systems. The six bridges of the direct frequency

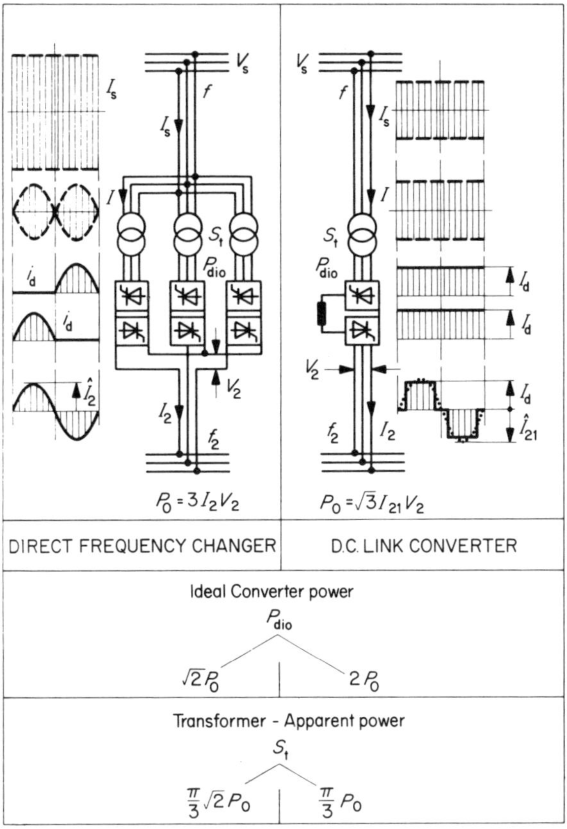

Figure 140. Both the three-phase direct frequency changer and the d.c.–link converter serve to translate power between two three-phase systems. Ideal comparison of the two converter variants. Direct frequency changer in ideal sinusoidal mode. Only the envelope is shown for the supply current (I_s). For derivations see Mathematical Appendices XVII and XVIII

changer, on the other hand, supply the phase voltage of the three-phase system. From the point of view of current, they are loaded only by half-waves of the low-frequency sinusoidal output.

The total *transformer power rating* in the direct frequency changer is greater than that in the d.c.-link converter. The transformer current in the direct frequency changer is modulated at low frequency, and therefore has a lower r.m.s. value than the constant transformer current in the d.c.-link converter, while the secondary windings of the transformer are dimensioned only in relation to the low-frequency phase voltages; nevertheless the total rating of the three transformers of the direct frequency changer is greater than that of the one transformer of the d.c.-link converter. At the same time, the d.c.-link converter requires the additional d.c. inductor.

MATHEMATICAL APPENDIX XVIII

The active powers P of the two three-phase systems (A and B) of the d.c.-link converter (Figure 136) are equal

$$P = \sqrt{3}\ V_A I_{A1} \cos \phi_{A1} = \sqrt{3}\ V_B I_{B1} \cos \phi_{B1}$$

and identical to the d.c. power

$$P = V_{dA} I_d = V_{dB} I_d$$

The fundamental components of the ideal rectilinear wave alternating currents of the inverters (Figure 85) are calculated as follows (equation 72):

$$I_{A1} = I_{B1} = \frac{\sqrt{6}}{\pi} I_d \tag{127}$$

The total r.m.s. values of the alternating currents are

$$I_A = I_B = \sqrt{\frac{2}{3}}\, I_d \tag{128}$$

The direct voltages are given by (equations (51), (67), and (74))

$$\begin{aligned} V_{dA} &= V_{dioA} \cos \phi_{A1} = \frac{3}{\pi} \sqrt{2}\ V_A \cos \phi_{A1} \\ V_{dB} &= V_{dioB} \cos \phi_{B1} = \frac{3}{\pi} \sqrt{2}\ V_B \cos \phi_{B1} \end{aligned} \tag{129}$$

The apparent power is

$$\begin{aligned} S_A &= \sqrt{3}\ I_A V_A = \frac{\pi}{3} V_{dioA} = \frac{\pi}{3} P_{dioA} \\ S_B &= \sqrt{3}\ I_B V_B = \frac{\pi}{3} I_d V_{dioB} = \frac{\pi}{3} P_{dioB} \end{aligned} \tag{130}$$

The total ideal d.c. power of both three-phase bridges is

$$P_{dio} = I_d V_{dioA} \left(1 + \frac{V_{dioB}}{V_{dioA}}\right) \tag{131}$$

or, if the two systems are similar,

$$\begin{aligned} V_{dioA} &= V_{dioB} = V_{dio} \\ P_{dio} &= 2 I_d V_{dio} \end{aligned} \tag{132}$$

Relative to the maximum ideal active power (the fundamental frequency apparent power)

$$P_0 = \sqrt{3}\ I_{21} V_2$$

the total ideal d.c. power is

$$P_{dio} = 2P_0$$

The rated transformer power (apparent power) of one system is

$$S_t = \frac{\pi}{3} P_0$$

SERIES CONVERTER CIRCUITS

General

A series connection of n individual three-phase bridges, each with a direct voltage V_{dio}, produces an overall terminal voltage equal to nV_{dio}. Thus if two converters are connected in series (Figures 118 and 119) the direct voltage is doubled.

The series connection of three-phase bridges is usually employed when the required direct voltage cannot be provided by a single bridge, because of the voltage loading of the valves. Obviously it is possible to connect two (or several) valves directly in series (see Figure 9), but the operation of the converter is not in that case modified, whereas with series-connected bridges *new converter characteristics* can be obtained.

Connecting two bridges in series enables the loading of the converter on the three-phase system—the harmonics or the reactive power—to be reduced.

Twelve-pulse converter circuit

The three-phase bridge provides a six-pulse supply to the load: in one cycle of the supply voltage six pulses are observed in the direct current and the direct voltage (Figures 29, 32, 37, 72, and 75). On the d.c. side the converter can be regarded as a harmonic voltage source of orders $6k$ ($k = 1,2, \ldots$). On the a.c. side the three-phase bridge functions as a harmonic current source of orders $6k \pm 1$ (see Figure 94).

The three-phase bridge can be supplied from a transformer with either star or delta connection (Figure 28), without its affecting the operation. In both connections the current in the primary winding has the same r.m.s. value and harmonic content. The current waveform is, however, different (Figures 115 and 141). The harmonic spectrum of the primary current is the same for the delta as for the star connection as regards amplitude (Figures 85 and 92), but the harmonics in the two circuits of orders 5, 7, 17, 19, etc., are in *opposition*. If two such currents are added, the harmonics that are in opposition are eliminated and the resulting current has a correspondingly *reduced harmonic content* (equation 95).

Figure 141 shows two three-phase bridges connected in series and supplied from transformers with star and delta connections respectively. *The resulting supply current contains no harmonics of orders 5, 7, 17, 19, etc.,* and its harmonic content is therefore only about half that of a single bridge.

The direct voltages of the two bridges are mutually displaced by 30°. Consequently the pulse number is doubled—the circuit functions *on the d.c. side in a twelve-pulse mode.* The sixth-harmonic voltage is eliminated.

The conductor and winding currents in the twelve-pulse converter circuit are illustrated in idealized form (smooth direct current, no overlap) in Figure 141.

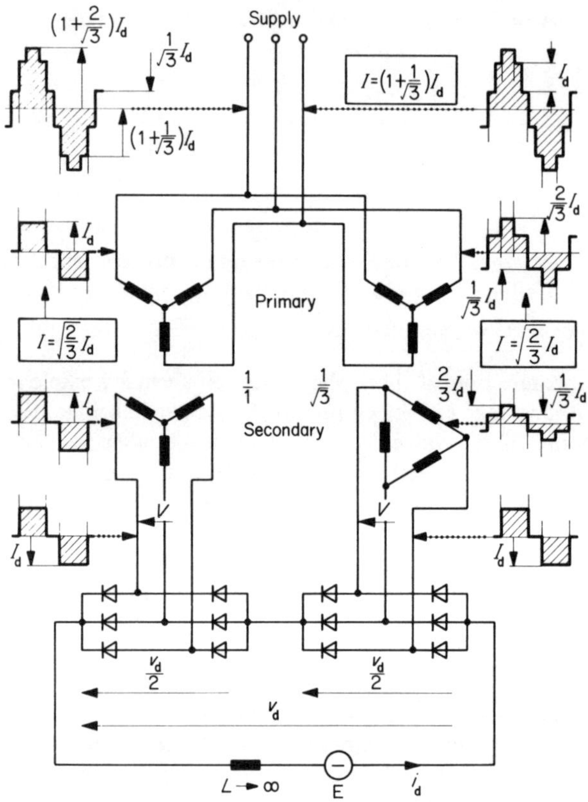

Figure 141. Series connection of two three-phase bridges with differently connected transformers (star and delta) leads to twelve-pulse converter operation. The harmonic content of the supply current is thereby reduced to about half. Illustration with idealized current waveforms

Examples of actual current waveforms with diode bridges can be seen in Figure 142. The corresponding amplitude spectrum of the currents is illustrated at the bottom of the diagram.

For a twelve-pulse converter it is usual to use one transformer with one primary and *two secondary windings.* The secondary windings are loosely coupled with one another magnetically, so that commutation notches (Figures 79 and 93) are not induced to an unacceptable degree in one winding from the other.

As well as the *series connection* of the bridges, twelve-pulse operation of the converter can also be achieved by means of *parallel-connected* bridges. An additional inductor has to be included in the d.c. circuit, since the direct voltages of the two bridges are not identical but only equal in mean value, and the difference voltage would give rise to an unacceptable *circulating current*.

The twelve-pulse converter bridge circuit is basically applicable to reversible converters, direct frequency changers, and d.c.-link converters, with the object

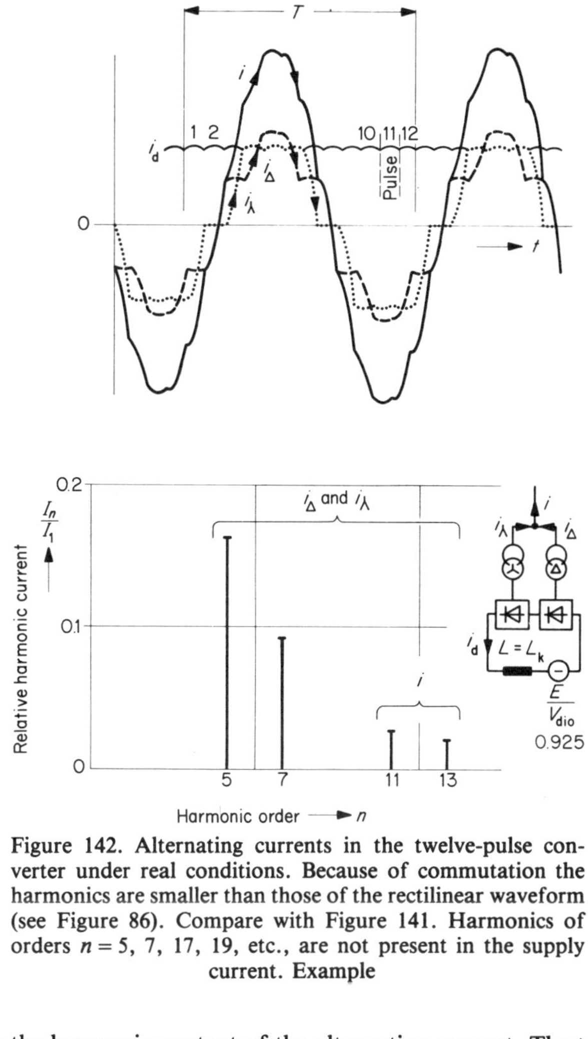

Figure 142. Alternating currents in the twelve-pulse converter under real conditions. Because of commutation the harmonics are smaller than those of the rectilinear waveform (see Figure 86). Compare with Figure 141. Harmonics of orders $n = 5$, 7, 17, 19, etc., are not present in the supply current. Example

of reducing the harmonic content of the alternating current. The twelve-pulse circuits are summarized in Figure 143.

In the series-bridge twelve-pulse circuit there are four valves in series in the current path. It is therefore necessary (in starting, or in discontinuous-current operation) to trigger four thyristors simultaneously. Since the three-phase systems of the two bridges are mutually displaced by 30°, four triggering pulses, separated by 30°, must be applied to each thyristor. (In the simple three-phase bridge each thyristor receives two pulses 60° apart—see Figures 48, 49, and 52). Control is effected by a common control system: all the triggering pulses are produced simultaneously, so that the delay angle is the same for both bridges.

The arrangement of the primary winding of the three-winding transformer for the twelve-pulse converter is a matter of choice, and has no effect on the operation of the converter.

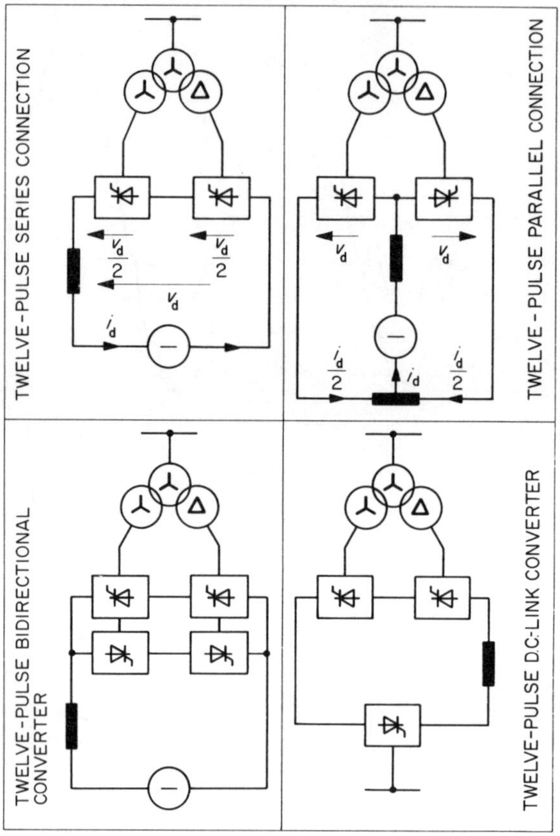

Figure 143. The transformer in a twelve-pulse converter normally has only one primary winding. The direct frequency changer and the d.c.-link converter can also be arranged for twelve-pulse operation. Twelve-pulse operation can be achieved by either series or parallel connection of two three-phase bridges

Converter with sequential control

The series connection of two similar three-phase bridges doubles the direct voltage. This, however, requires either two similar transformers (Figure 141) or a transformer with two secondary windings of equal voltage (Figure 143).

If the two bridges are triggered simultaneously by a single control unit ($\alpha_A = \alpha_B$), the converter has either six-pulse characteristics, when the transformer windings are identical, or twelve-pulse, when the transformer has star-delta windings (Figures 141 and 142). In both cases the individual direct bridge voltages are each half the total voltage under all conditions.

Two similar three-phase bridges in a series circuit can, however, be *separately controlled.* Since each bridge is capable of producing a direct voltage of $\pm\frac{1}{2} V_d$, the condition of zero converter voltage, $V_d = 0$, can be achieved by two opposing part-voltages, $+\frac{1}{2} V_d$ (full rectifier output, $\alpha \rightarrow 0°$) and $-\frac{1}{2} V_d$

(full inverter output, $\alpha \rightarrow 180°$). The converter can then be controlled in such a way that only one bridge is fully controlled, while the second bridge remains in the condition of maximum rectifier or inverter output.

This sequential control of the converter thus has the effect that only half the active power ($\frac{1}{2}\ V_d I_d$) is controlled at any time. The reactive power diagram of the converter with sequential control consists therefore not of one semicircle (Figures 73 and 77) but of two semicircles, as shown in Figure 144. In comparison with simple control, with sequential control the fundamental frequency reactive power, for a given active power, is considerably smaller. The difference is particularly large for low direct voltages ($v_d = 0$).

Sequential converter control saves reactive power, particularly in high-power applications, when starting is frequent (as in reversing d.c. drives; see Figure 121) or when low output is maintained for long periods (delay angle around 90°).

As well as reducing the reactive power, sequential converter control also reduces the supply current harmonics. Depending upon the operating point, the harmonic content lies between the values corresponding to six-pulse and twelve-pulse operation (Figure 144, right).

Figure 118 indicates the place of sequential control in the classification of double-bridge converter circuits. Like the twelve-pulse circuit, sequential converter control can be applied to reversible converters, direct frequency changers, and d.c.-link converters (Figure 143). The transformer arrangement

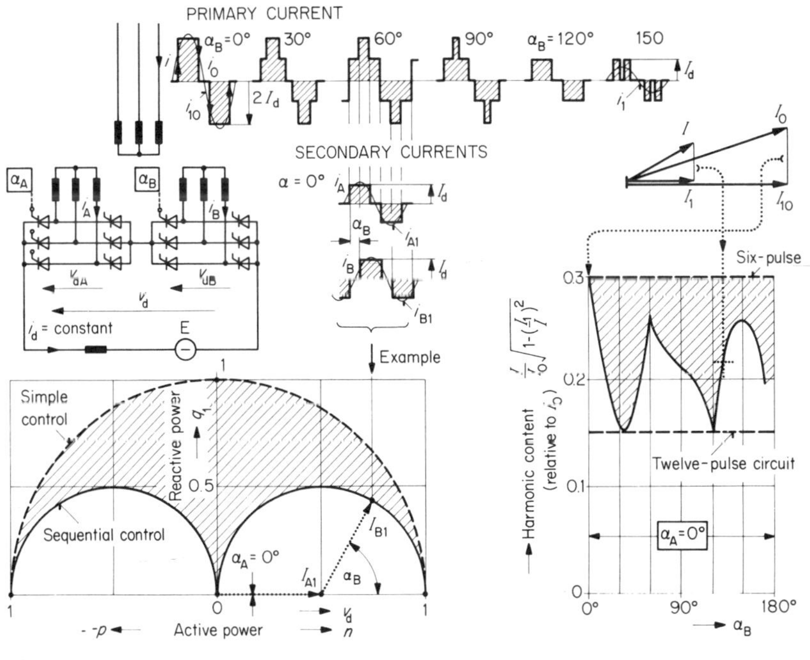

Figure 144. By appropriate sequential control of two three-phase bridges connected in series the control reactive power and the harmonic content (equation 94) of the converter can be reduced. Idealized presentation (smooth direct current, no overlap). I_0 = r.m.s. value with $\alpha = 0°$; I_1 = fundamental component

has no significant effect. With a star–star connection (Figure 144) the harmonic content corresponds at maximum output ($\alpha = 0^\circ$) to that of the six-pulse converter, and with $V_d = 0$ to that of the twelve-pulse converter. With star–delta secondary windings the situation is reversed. The primary winding arrangement of the three-winding transformer is open to choice, and has no effect on the operation of the converter.

13

Converter–Supply Interactions

GENERAL

The three-phase bridge is connected either to the three-phase supply system (through a transformer) or to a three-phase machine. The three-phase voltage system is essential to the operation of the converter. The converter valves carry the current cyclically in sequence which is determined by the three-phase system (Figures 29 to 32). For commutation the converter needs reactive power from the three-phase system.

During the current transfer (commutation) two or three phases of the three-phase system are briefly short-circuited (Figures 38 to 41). Voltage drops are developed across the commutating inductances (Figure 79) and the alternating current contains harmonics (Figures 85 to 93). From the point of view of the three-phase system the converter represents a series of *harmonic current sources* (Figure 94).

The load characteristics of the three-phase bridge (for a given load) are determined by the parameters of the three-phase system (Figures 112, 116, and 117). In principle the converter is merely a *switching arrangement,* which connects the three-phase system to the load system (d.c. or a.c.) is a specified manner (Figures 31, 32, and 40). If the supply parameters change, so do the converter characteristics. Between the converter and the three-phase system there exists a close interaction.

LOADING OF THE THREE-PHASE SYSTEM BY THE CONVERTER

The conventional theory of the three-phase bridge is based upon the assumption of an ideal inductive three-phase system (Figure 70). This system is loaded by the operation of the converter with *reactive power* (Figures 73 and 77) and *harmonics* (Figures 85 to 97). The power-related parameters of the converter—power factor, displacement factor, harmonic content, and harmonic spectrum—are dependent upon the loading, control, and configuration of the converter (for the direct frequency changer see Figures 126 to 130, for the twelve-pulse converter see Figure 142, and for sequential control see Figure 144).

During commutation (μ) the line voltage at the a.c. terminals of the converter is brought to zero (Figure 93). At the point of connection of the converter transformer the supply voltage v_s drops momentarily according to the ratio of the supply and transformer short-circuit inductances (L_s/L_T). Figure 145 shows an example of the distortion of the supply line voltage, the current waveform, and the corresponding harmonic spectra.

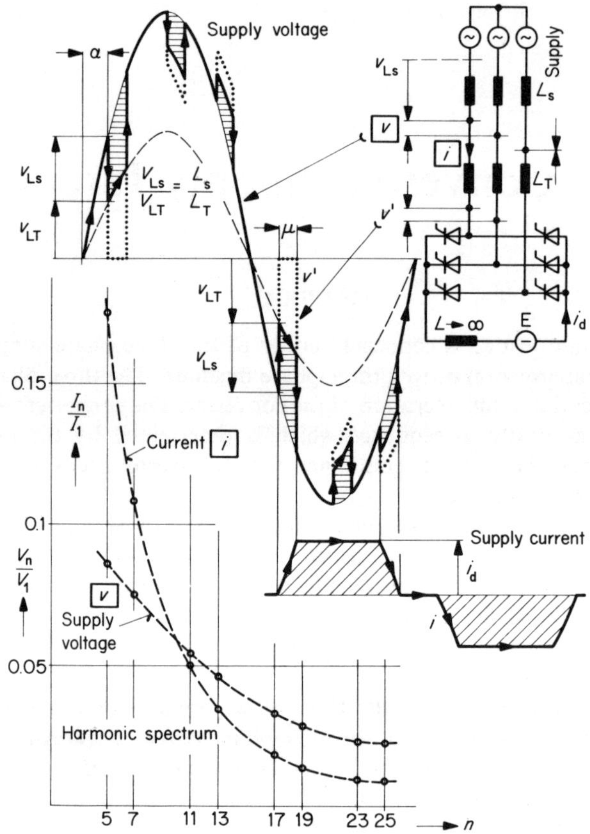

Figure 145. As a result of commutation μ in the converter the supply voltage v is distorted. The notches in the supply voltage depend upon the ratio between the supply and transformer short-circuit inductances (L_s/L_T). Example with ideal inductive supply and smooth direct current

The short-circuit inductances of the transformer and the supply, L_T and L_s, are inversely proportional to their short-circuit VA (Q_T, Q_s; see also Figures 116 and 117). The higher the supply short-circuit level (relative to the transformer short circuit VA) the less is the supply-voltage distortion. The harmonic content of the supply voltage is greatest when the delay angle α is 90° (for a given direct current). Figure 146 illustrates the relationship of the harmonic content k_H of the supply voltage to the short-circuit VA ratio Q_s/Q_T.

If several converters are connected to the same three-phase system (supply mains) they will affect one another through the commutation notches. Operation directly in parallel is (theoretically) not possible. During the overlap period μ of one converter the line voltage is zero and the thyristors of the other converter cannot be triggered. The converters must therefore be *decoupled* by inductances. This is shown in Figure 147. The decoupling inductances L_T are so chosen that (with a minimum delay angle α_{min}) the supply voltage during

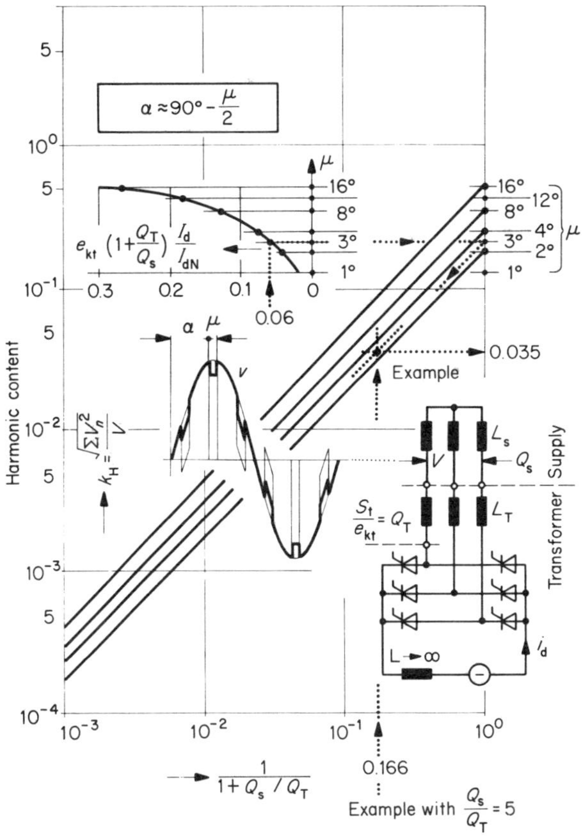

Figure 146. The harmonic content of the supply voltage, distorted by the converter harmonics, depends upon the ratio between the supply and transformer short-circuit volt-amperes. Graphical illustration with an unfavourable control situation, purely inductive supply, and smooth direct current

the commutation period μ is not below the level necessary for certain triggering (ν_{Thmin}).

Actual supply systems are not purely inductive, but include capacitances as well as ohmic resistances. They are therefore oscillatory (see Figure 19). The *supply capacitances,* together with the system inductances, form oscillatory circuits which are excited by the commutation voltage steps. This is illustrated in Figure 148. If the supply is considered as including capacitances *C*, oscillations corresponding to the natural frequency f_0 of the supply system are superimposed on the supply voltage. These oscillations are clearly discernible in the amplitude spectrum of the harmonics. For comparison, Figure 148 also shows the harmonic spectrum of the supply voltage with a purely inductive system.

The supply current harmonics of the converter (Figure 85) can be considered in terms of phasors. When several converters are connected to one supply

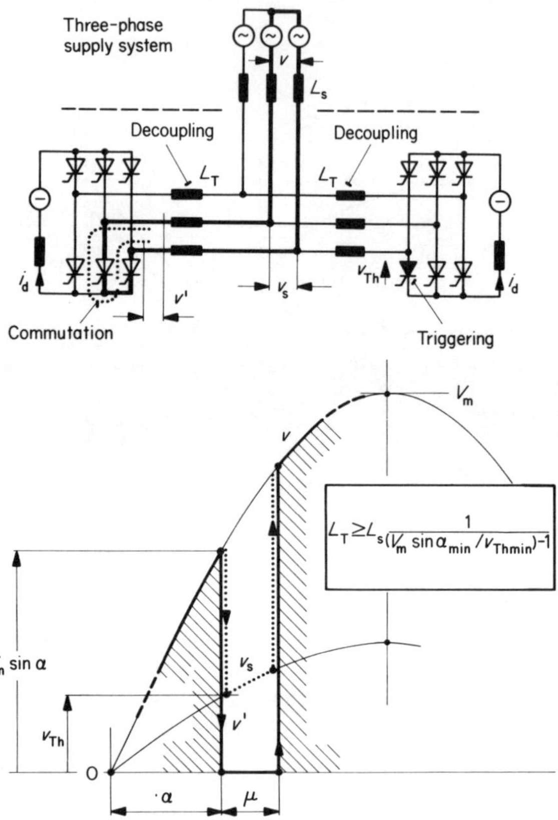

Figure 147. Two (or more) converters on one three-phase supply must be decoupled by inductances (L_T). When one converter commutates, a sufficient voltage v_{Thmin} must be available to ensure triggering of the thyristors of the other converter. (Standard limits: $\alpha_{min} = 10\,°$, $v_{Thmin} = 10$ V.)

system, individual harmonics of the same order, generated by the converters, are added geometrically. The highest possible resultant is represented by the algebraic sum. At a minimum, the harmonics may cancel out completely.

Figure 149 shows, for example, the way in which the eleventh and thirteenth harmonics due to two similar converters vary as one converter (A) is controlled (converter B remaining at $\alpha = 0°$).

Depending upon the delay angle α_A the supply harmonics lie between the maximum of approximately the algebraic sum and a minimum value. The extreme values of the two harmonics do not, however, occur at the same time.

When a number of converters are connected to one supply on the assumption that they are (statistically) independent units it is appropriate to assess their overall behaviour in regard to harmonics *statistically*. The harmonics can be considered as random phasors, representing a random magnitude and a random phase angle. It is thus possible to assign a *probability* to the

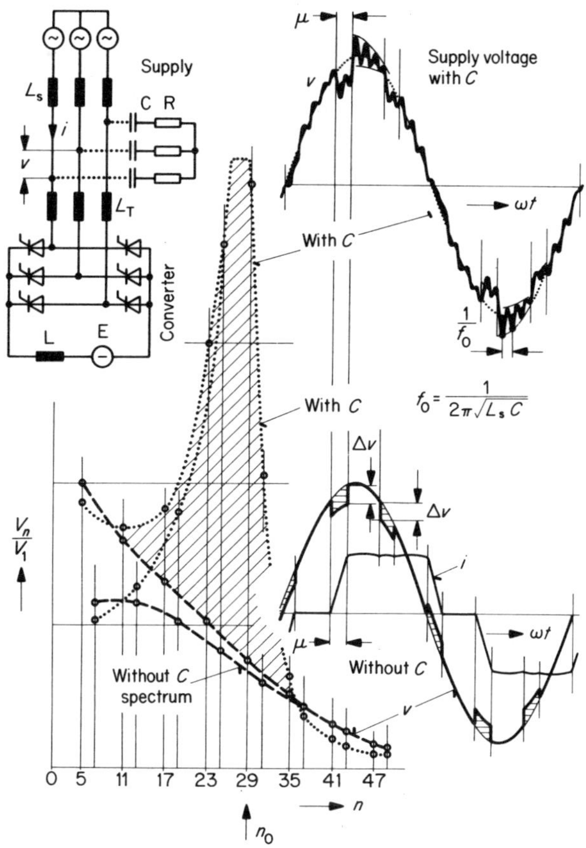

Figure 148. Real three-phase supply systems often contain capacitances (C) which can form oscillatory circuits with the supply inductances (L_s). Commutation in the converter produces steps (Δv) in the supply voltage which excite these oscillatory circuits. The supply voltage is then further distorted by ringing. Harmonic spectrum (the linking curves have no physical significance), voltage waveform, and diagram for calculation. Example

geometrical sum of these random phasors—the phasor resultant—for any value between zero and the absolute maximum. The greater the number of units (i.e. of random phasors) the more improbable is it that the algebraic sum will be reached.

Figure 150 illustrates the reduction in the magnitude of the phasor resultant, relative to the algebraic sum, as a function of the number of converter units for two particular probabilities. It will be seen that with ten units, for example, it can be assumed with 99.9 per cent confidence that the phasor resultant of the harmonics will not exceed half (50 per cent.) of the algebraic maximum. (The curves in Figure 150 are based on the assumption that the absolute values of the phasors are randomly distributed between 0 and 1 and that their phase angles lie randomly between 0° and 360°.)

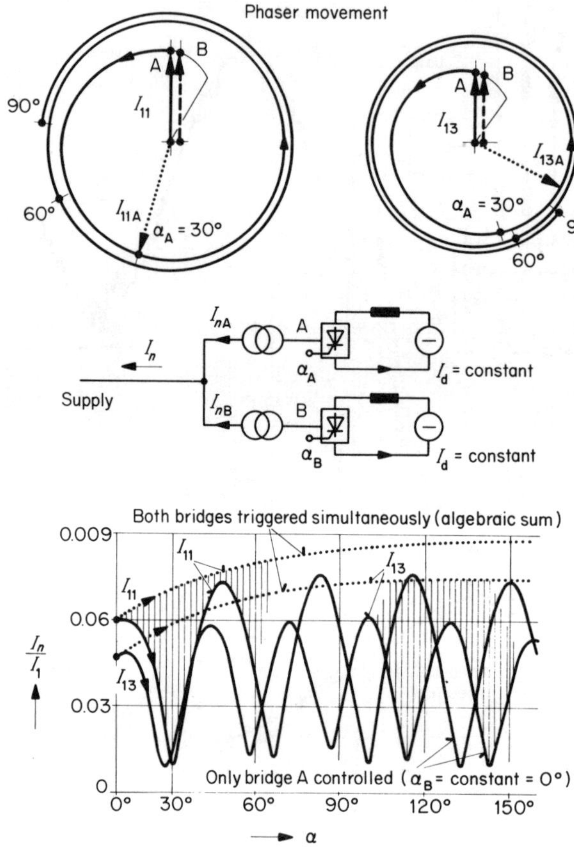

Figure 149. The harmonics of two (or more) converters connected to the supply add geometrically. Example with two converters similarly loaded but differently controlled. Phasors and amplitudes of the eleventh and thirteenth harmonics with $dx = 0.04$. Smooth direct current

MATHEMATICAL APPENDIX XIX

In simple commutation two phases of the three-phase system are short-circuited. The source voltage is divided in the short-circuit path according to the inductances (see Figure 147) as follows:

$$\frac{V_s}{V} = \frac{L_T}{L_T + L_s} = \frac{1}{1 + L_s/L_T}$$

To ensure at least a minimum supply voltage $v_s > v_{\text{Thmin}}$, the decoupling inductance (for two converters in accordance with Figure 147) must be

$$L_T \geqslant L_s \frac{1}{(v/v_{\text{Thmin}}) - 1}$$

Here the source voltage v is that at the beginning of commutation (delay angle

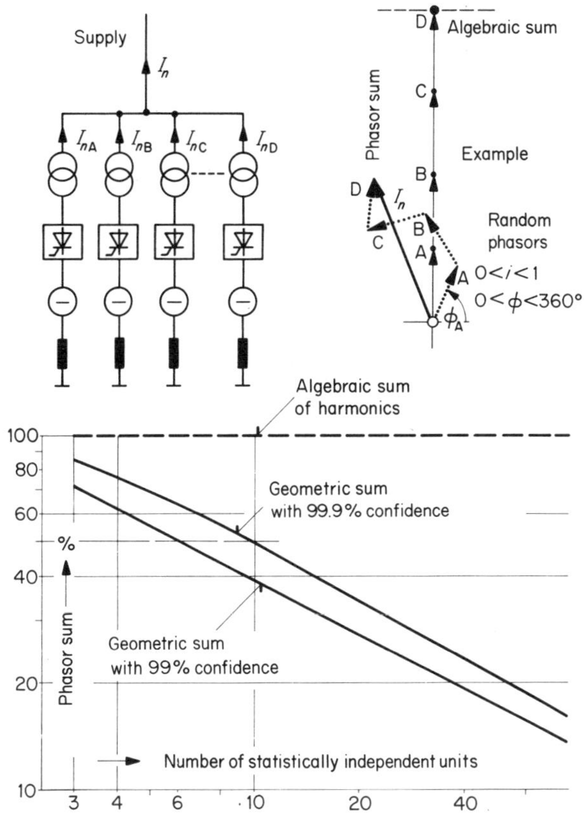

Figure 150. When several independent converters are connected to a single supply, statistical considerations can be applied to the estimation of the total supply harmonics. The harmonics combine statistically and, to a high degree of confidence, will not exceed the calculated level indicated by the graph (the result of a computer simulation study, using a Monte Carlo method). It is assumed that fluctuations in amplitudes and phase angles are statistically evenly distributed

α). Certain thyristor triggering is required at some angle $\alpha_{min} > 0^\circ$. Then

$$v = V_m \sin \alpha_{min}$$

and

$$L_T \geqslant L_s \frac{1}{(V_m \sin \alpha_{min}/v_{Thmin}) - 1} \tag{133}$$

If there are $n' > 2$ converters connected to the supply, assuming that $n' - 1$ commutate simultaneously and that all n' inductances L_T are equal,

$$\frac{v_s}{v} = \frac{1}{1 + (L_s/L_T)(n' - 1)}$$

and

$$L_T \geqslant \frac{L_s(n'-1)}{(V_m \sin \alpha_{min}/v_{Thmin})-1} \tag{134}$$

COMPENSATION FOR CONVERTER SUPPLY LOADING

Both the fundamental frequency reactive power and the harmonic non-active power of the converter can be compensated by inductive-capactive filter circuits, so that the three-phase system is relieved.

Since the three-phase bridge generates current harmonics only of certain orders ($n = 5, 7, 11, 13, \ldots$), these can be filtered out by simple tuned *LC series-resonant circuits* connected to the supply in parallel with the converter.

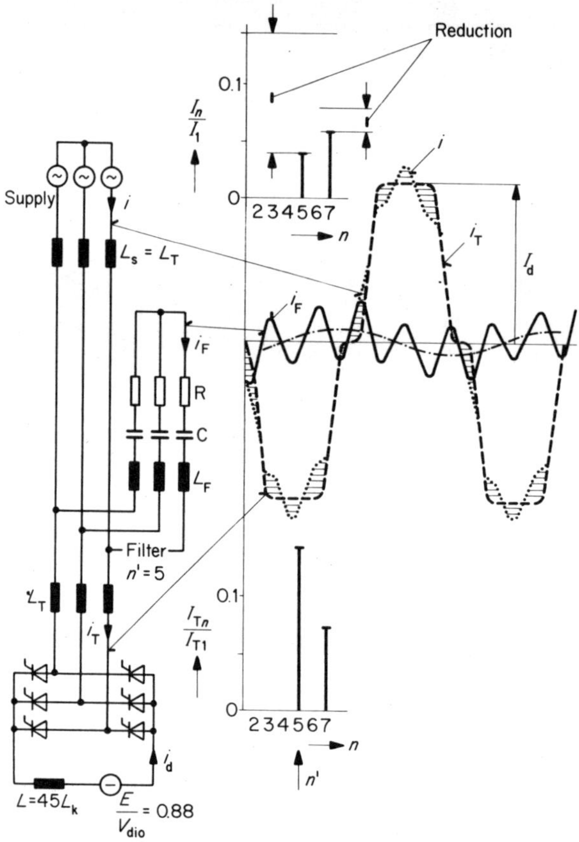

Figure 151. A series-resonant circuit (CL_F) tuned to the frequency of a converter harmonic (n') acts as a filter and so reduces the corresponding current harmonic in the supply. Example for $n' = 5$. At the same time the filter compensates for the fundamental frequency reactive power. The chain-dotted mean curve through i_F represents the fundamental component

These filters are capacitive at the fundamental frequency; i.e. they compensate simultaneously for the fundamental frequency reactive power as well.

In the operation of the filter, the ohmic resistance of the series resonant circuit (the 'Q' factor) is of special significance. The smaller the ohmic resistance, the more effective is the filter (i.e. the greater is the reduction in the supply current harmonic), but only under steady-state conditions. Under dynamic conditions—load variations, etc.—a very low resistance has an adverse effect on the filtering performance (an ideal zero-resistance filter would have no effect at all).

The higher the supply inductance (the lower the supply short-circuit level), the more effective is a given filter. A particular filter is therefore more effective connected directly across the a.c. side of the converter (the secondary side of

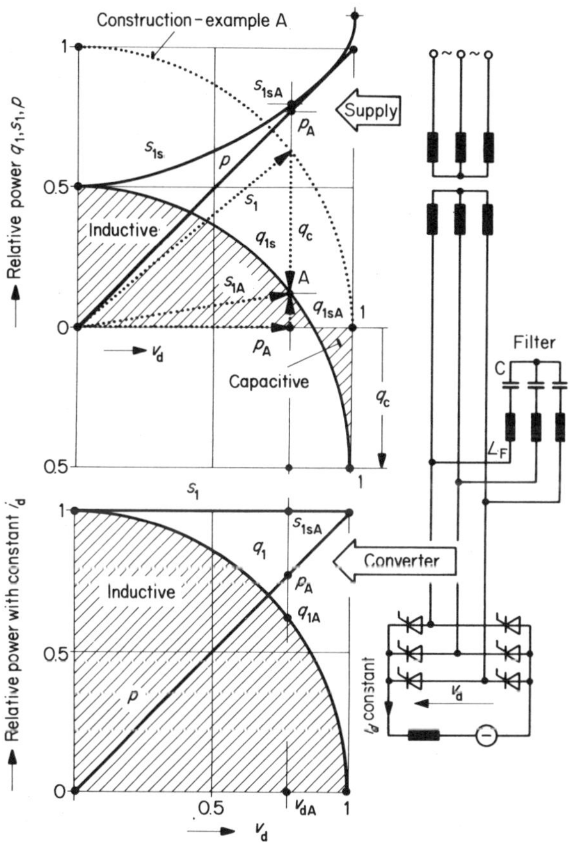

Figure 152. The converter loads the supply with a lagging fundamental frequency reactive power (q_1) which can however be compensated by a capacitance q_C. Example of the compensating effect of a filter circuit. Power circle diagram, in relative values, for the converter (bottom) and the compensated supply (top). Dotted: auxiliary construction for example A

the transformer) than across the supply (the transformer primary). The voltage stress on the valves is thereby increased, however.

A filter (series-resonant circuit) for the *n*th harmonic forms with the supply inductance a parallel-resonant circuit whose natural frequency is less than that of the filter. The filter assembly must therefore be built up from the lowest harmonic order ($n = 5$) and tuned to it. If, for example, only one filter for the eleventh harmonic is connected, that harmonic is reduced, but the seventh and fifth are increased.

If it is only required to compensate for the fundamental frequency reactive power, an LC circuit should still be used, tuned to a natural frequency lower than that of the lowest converter harmonic ($1 < n < 5$).

An example of the effect of a harmonic filter ($n = 5$) is illustrated in Figure 151. This shows the current waveforms in the converter and the supply, and also their amplitude spectra, from which it can be seen how effectively the harmonic is reduced by the filter. It can also be seen from the filter current i_F that, as well as the prominent fifth harmonic, there is present in addition a leading fundamental component: i.e. *the filter also presents a capacitance* and provides reactive compensation.

Compensation of the fundamental frequency reactive power by an LC filter circuit is illustrated diagrammatically in Figure 152. The power balance in the converter and the supply is shown by the circle diagram (as a function of v_d) for the whole control range in the rectifier mode, and a direct current i_d for an arbitrary compensating *VA* q_c. (The dotted lines above are construction lines and are comparable with Figure 76.)

MATHEMATICAL APPENDIX XX

The natural frequency of an LC series-resonant circuit (Figure 19, equation 14) is equal to

$$f_0 = \frac{1}{2\pi\sqrt{LC}} \tag{135}$$

or

$$\omega_0 = \frac{1}{\sqrt{LC}}$$

At this frequency the resonant circuit represents a zero resistance, and thereby functions as a filter. To keep the (50 Hz) supply mains free from harmonics (with three-phase bridge loading; see Figure 151) the filter frequencies must be

$$f_0 = 250\ \text{Hz},\ 350\ \text{Hz},\ 550\ \text{Hz},\ 650\ \text{Hz},\ 750\ \text{Hz, etc.}$$

The reduction in the harmonic current I_n generated by the converter due to the filter (of the same natural frequency) is given by the equation (steady-state)

$$\frac{I_{ns}}{I_n} = \frac{1}{\sqrt{1 + (\omega_0 L_s/R)^2}} \tag{136}$$

where L_s = supply inductance
R = resistance of filter

I_n = harmonic current from converter
I_{ns} = harmonic current in supply
$\omega_0 = 1/(\sqrt{L_F C})$

The fundamental frequency current in the filter can be expressed as

$$I_1 = \frac{V_{ph}}{\omega L - 1/(\omega C)} = \frac{V_{ph}}{\omega L} \frac{1}{n^2 - 1}$$

$$= V_{ph}\,\omega C \frac{n^2}{n^2 - 1} \tag{137}$$

where ω = angular frequency of supply voltage
$n = \omega_0/\omega$ = order of harmonic

The capacitive fundamental frequency reactive power of the filter is therefore (for each LC circuit)

$$Q_C = V_{ph}^2\,\omega C \frac{n^2}{n^2 - 1} \tag{138}$$

and for all three phases

$$Q_C = V^2\,\omega C \frac{n^2}{n^2 - 1}$$

THE EFFECT OF THE SUPPLY ON THE CONVERTER

One and the same converter (with its matching transformer) will exhibit different load characteristics (Figures 116 and 117), harmonics, and power relationships on supplies with different short-circuit levels. The supply-voltage notches produced by converters operating in parallel (Figure 147) are coupled to the secondary side of the transformer, and thence to the thyristor voltage. This can cause, on the one hand, triggering problems and, on the other hand, loss of inverter control if the voltage dip reduces the safety margin γ (see Figure 62) so that the outgoing thyristor cannot turn off.

Particularly critical is the mutual influence of converters that are supplied from a multi-winding transformer (converters with sequential control, three-phase direct frequency changers, etc.). The design of the transformer should be such that the secondary windings are as loosely coupled as possible.

Reactive power and harmonic compensation is usually effected in converter installations by capacitor banks or filters (Figures 151 and 152). *The a.c.-side capacitances C reduce the commutating times of the converter* and thereby alter the load characteristics.

The compensating circuits are most effective when connected directly to the a.c. terminals of the converter. This is shown as an example in Figure 153, which shows a filter for the suppression of the fifth harmonic connected directly to the converter terminals. The load characteristic is in consequence altered (it becomes 'stiffer') and the harmonic content of the supply current is reduced (most markedly, the fifth harmonic). The harmonic content of the alternating current in the converter is, however, increased. At the same time the action of the filter increases the voltage on the valves.

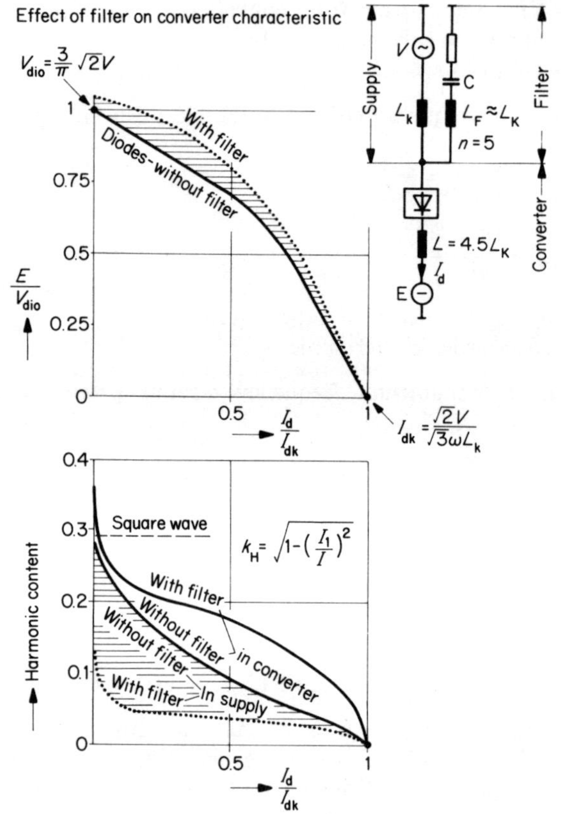

Figure 153. The supply current harmonics produced by the converter can be effectively suppressed by filters. The filters, however, react upon the converter and modify its characteristics. An arbitrary example with a directly connected filter for the fifth harmonic

14

Converter Protection

PROTECTION PHILOSOPHY

By the term 'converter protection' is normally understood *the protection of converter valves.* Converter valves, thyristors or diodes, can be destroyed by voltages in excess of the voltage rating of the valves or by currents that cause excessive heating in the valves.

If the negative blocking voltage of a semiconductor valve exceeds a critical value, the breakdown voltage (Figure 5), the result is an electrical puncture, which destroys the internal p–n structure of the silicon wafer (Figure 4). The punctured valve behaves thereafter like a low-resistance conductor. The breakdown voltage in the reverse direction is *temperature-dependent.* At very high or very low temperatures (above about 150 °C or below about − 50 °C.) the blocking voltage capability of the valve is reduced.

If the positive blocking voltage across the thyristor exceeds, even momentarily, the breakover voltage (Figure 5), the thyristor turns on (without a triggering pulse), and this too can lead to destruction of the internal structure. The breakover voltage is also temperature-dependent: at temperatures above about 125 °C it falls sharply.

Under normal operating conditions the voltage stressing of the valves in the three-phase bridge is related to the peak line voltage V_m of the three-phase system (Figures 30, 49, 50, 55, and 57). The valve current is determined by the delay angle and the d.c. circuit (Figures 57, 58, and 64).

Under a converter fault condition (Figures 99 to 102, 105 to 111, and 139), valves carry currents many times their rated currents and are thereby endangered.

Overvoltages, which exceed the nominal voltage loading of the valves (V_m), arise firstly in the converter itself, from the non-ideal operation of the semiconductor elements, and secondly from transient phenomena in the a.c. or d.c. circuits. Valves are normally rated, from economic considerations, with a voltage safety factor of two. Any overvoltages must be suppressed to within this value.

It is relatively easy to calculate the loading of the converter valves through short-circuit currents (see Figure 111). The problem of estimating overvoltages, which are generally stochastic in nature, is more difficult.

Apart from overcurrents and overvoltages, thyristors are also at risk from very sharply rising forward voltage steps (dv/dt), which can lead to unintended turning-on without triggering pulses, and high rates of rise of forward current (di/dt), which can cause breakdown on the thyristor through local overheating of the silicon wafer.

The most important protective measures against overvoltages and overcurrents will be discussed further. The application of practical protection devices is, however, governed by the 'protection philosophy'. Such considerations take account of the following factors:

The probability of faults
Consequential damages
The importance of the installation
Experience, etc.

The choice of protection system is also concerned with the required fault indications and with redundancy.

The fault liability of an installation generally increases with the number of components in it. Thus the protection devices have no positive function but increase the overall probability of failure. In addition, the protection devices, through their *secondary effects,* often endanger the protected components in

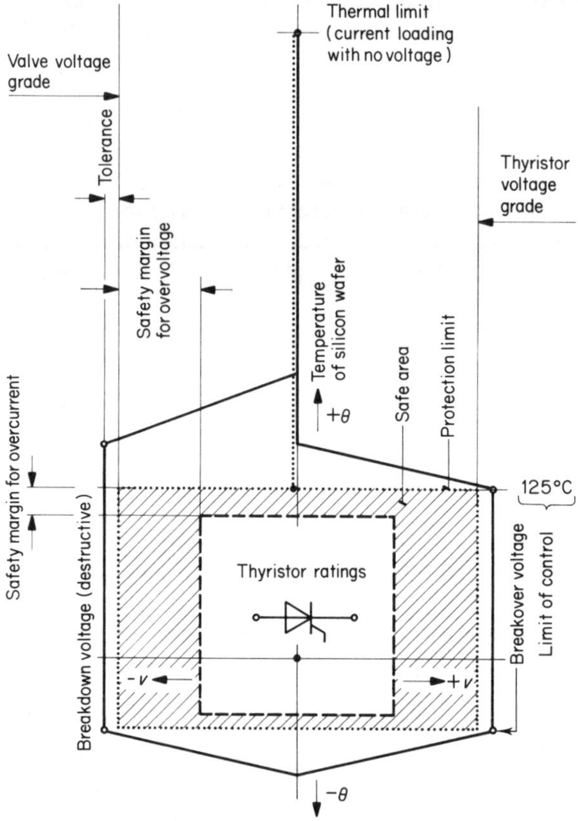

Figure 154. The electrical stability of the thyristor is temperature-dependent. If the limit values are exceeded the thyristor may be destroyed. The protection devices ensure that these limits are not exceeded in operation (through overvoltages or overcurrents). Diagrammatic representation

another way. So, for example, the circuit breaker and the fuse, which protect the converter valves against overcurrent, generate additional overvoltages through their switching action.

Figure 154 shows diagrammatically, on voltage and temperature axes, the protection area and the electrothermal limits of the thyristor.

OVERVOLTAGE PROTECTION OF CONVERTER VALVES

Overvoltages, general

By overvoltage is meant a voltage that exceeds the nominal value of the supply voltage (the maximum, or peak, value). The physical origin of the overvoltage lies either in rapid *variations of current (*$\mathrm{d}i/\mathrm{d}t$*) in circuit inductances* or in *the charging of circuit capacitances.*

Overvoltages are produced by *sudden changes of state* in switching circuits, such as switching on and off, current and voltage steps, or periodic excitation of resonant circuits.

In three-phase rotating machines, overvoltages arise as a result of relative variations of the rotating magnetic field. From the point of view of the voltage stressing of the valves in a converter, the full operating cycle of the converter can be divided into the following periods:

Switching of the converter onto the three-phase system
Discontinuous current operation of the converter
Normal operating condition with commutation
Switching-off of the converter

Switching operations and faults mostly produce single unrepeated overvoltages of short duration. During the operation of the converter, limited periodic overvoltages are produced by commutation.

Apart from the *semiconductor valves,* the structural parts of the converter, the *insulation* between the bridge arms and the insulation between the power circuits and the auxiliary and control circuits and earth are also subjected to voltage stress.

The converter valves are usually specified with a voltage safety factor of two (relative to the peak line voltage; see Figure 57), so the protection devices must limit overvoltages to less than this value.

The *capacitor* may be considered the most important protection component in converter engineering. In a capacitance the surplus energy, which would otherwise give rise to an overvoltage, can be absorbed at a low voltage level. Subsequently, after the event (e.g. the switching operation) is over, the stored energy is dissipated in a discharge resistor and converted into heat.

Converter overvoltage protection circuits generally embody capacitors. These, however, form *oscillatory circuits* in combination with circuit inductances, and this can lead to undesirable secondary effects.

A voltage step applied to an undamped inductive–capacitive circuit charges the capacitor to twice the input voltage (Figure 19).

In addition, discharge currents from the capacitors in some circumstances constitute a danger to the thyristors through high rates of rise ($\mathrm{d}i/\mathrm{d}t$).

Switching-on of the converter

In a blocking three-phase bridge with no current (without triggering pulses, or with $\alpha > 120°$), the thyristors are subjected under steady-state conditions to the phase voltage of the three-phase system (Figure 48).

When the converter transformer is switched onto the supply, undesirable effects can be caused by the *transformer winding capacitances.* During the switching-on process (the voltage step) the primary voltage can be coupled directly to the secondary side through the capacitance (not according to the transformation ratio). The insulation of the converter valve circuits from earth is particularly stressed.

A stationary *induction motor* behaves like a transformer. When the stator winding is connected to the supply a voltage appears on the rotor side in proportion to the turns ratio. This has to be taken into account in the voltage specifications for the subsynchronous converter cascade (Figures 118 and 119).

Actual *three-phase switches* do not switch all three phases of the system

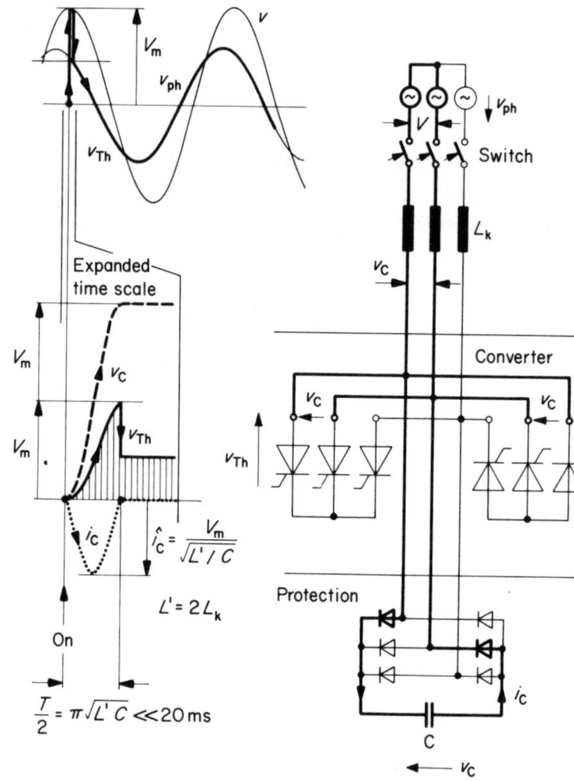

Figure 155. Overvoltage protection is usually provided by a capacitor connected to the three-phase terminals through diodes. At switch-on the capacitor voltage v_C can rise to twice the peak supply line voltage $2\sqrt{2}\,V$. Voltage relationships with a non-conducting three-phase bridge. Undamped

simultaneously, but with a time displacement. Mechanical switch contacts can also bounce after closing. These two features are often associated with overvoltages.

The *supply system capacitances* form oscillatory circuits with the supply and transformer inductances, and can give rise to overvoltages when the circuit is closed. The inrush current of the transformer has a large harmonic content. The harmonic currents during the switching-on period can provoke resonance in the inductive–capacitive supply circuits, and thus lead to an increase in voltage.

Supply overvoltages are normally limited to a known voltage level by *surge suppressors.* To prevent the capacitive coupling of voltages from the primary to the secondary side of the transformer, an earthed electrostatic screen is incorporated between the windings.

For a.c.-side overvoltage protection of the converter a capacitor is often connected through a diode bridge (Figure 155). At switch-on the capacitance C forms an oscillatory circuit with the total circuit inductance L', and the capacitor voltage v_C rises to approximately twice the input voltage (Figure 19). Switching on at the peak of the line voltage $\sqrt{2}\ V = V_m$ therefore produces a capacitor voltage of $2\sqrt{2}\ V$ (assuming that the natural frequency of the oscillatory circuit is much higher than the supply frequency). Only half the voltage is applied to the blocking thyristors, however (assuming equal voltage distribution). This is the situation during the charging period of the capacitor, when the diodes are conducting and the capacitor voltage is applied directly to the thyristors. When the capacitor current i_C falls to zero, the diodes block and the phase voltage appears across the thyristors (see Figure 155). The diodes are subjected to the full capacitor voltage during the charging process.

The voltage overshoot can be damped by a *resistor* in series with the capacitor.

Sustaining load for discontinuous current operation

Upon the triggering of the thyristor by a current pulse i_G conveyed to the gate by the pulse coupler (Figure 8), the thyristor voltage collapses and the current rises. The gate pulse must last long enough for the *latching current* of the thyristor to be reached, otherwise the thyristor will not remain in the conducting state but will turn off (Figures 5 and 6).

With a high ratio of load to commutating inductance ($L \gg L_k$) the discontinuous current is very small (Figures 65, 66, and 67) and the rise of current is therefore very slow, so that the latching current may not be reached during the triggering period. This can cause *triggering problems,* and the converter may not be controllable at low currents. The current steps produced by the turning-off of the thyristors can generate overvoltages. To avoid this a *sustaining load* consisting of a resistor and a capacitor is connected in parallel with the load inductance. This sustaining load ensures that the latching current of the thyristor is attained under any operating condition.

The functioning of the sustaining load is illustrated by the example of Figure 156. The rise of current in the thyristors (di/dt) is greatly increased by the sustaining load.

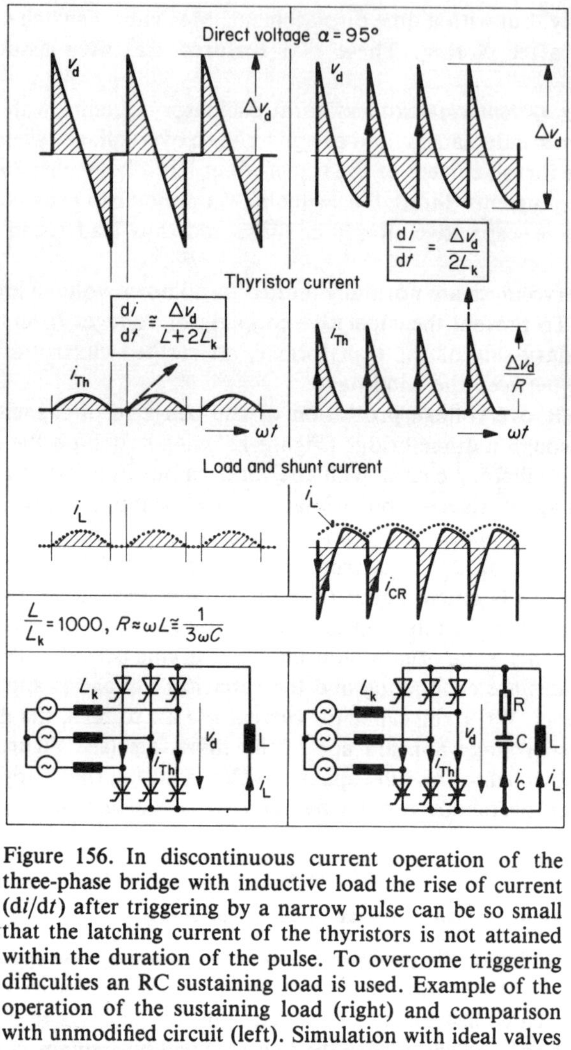

Figure 156. In discontinuous current operation of the three-phase bridge with inductive load the rise of current ($\mathrm{d}i/\mathrm{d}t$) after triggering by a narrow pulse can be so small that the latching current of the thyristors is not attained within the duration of the pulse. To overcome triggering difficulties an RC sustaining load is used. Example of the operation of the sustaining load (right) and comparison with unmodified circuit (left). Simulation with ideal valves

If there are capacitors in the converter (*individual RC suppression,* for example; see Figures 8 and 9), these can cause additional forced turn-off of the thyristors in discontinuous current operation, due to oscillatory currents.

In the *three-phase direct frequency changer* (Figures 127 and 131), notches in the secondary-side voltage due to commutation in the other converter phases can also lead to forced turn-off of the thyristors. In addition, a too-steep waveform in *trapezoidal operation* of the direct frequency changer with a synchronous machine (Figure 132) produces overvoltages.

In these situations also a suitable sustaining load serves as a remedial measure.

Commutation overvoltage in normal operation

Operation with simple commutation (Figure 71) represents the typical operating condition of the three-phase bridge.

During the commutating period (μ) the thyristor current rises and falls very rapidly (Figures 38, 57, and 75). The commutating current corresponds to the short-circuit current i_k of the three-phase system. The highest rate of rise of current due to commutation occurs with a delay angle $\alpha = 90^\circ$.

It can be seen from the dynamic characteristics of an actual thyristor (Figure 5) that the transition from the conducting to the reverse blocking state is accompanied by a *reverse current spike*. The thyistor does not block instantaneously when the forward current falls to zero, but continues to conduct freely in the reverse direction for a certain time (Figure 6). Only when the reverse current has reached a particular value is it terminated sharply. *The thyristor thus turns off with a delay.*

The phenomenon is referred to as the *carrier storage effect*. During this period the semiconductor layers of the thyristor are emptied of charge carriers (Figure 6).

Thus the thyristor, assumed for the purposes of circuit theory to be a perfect

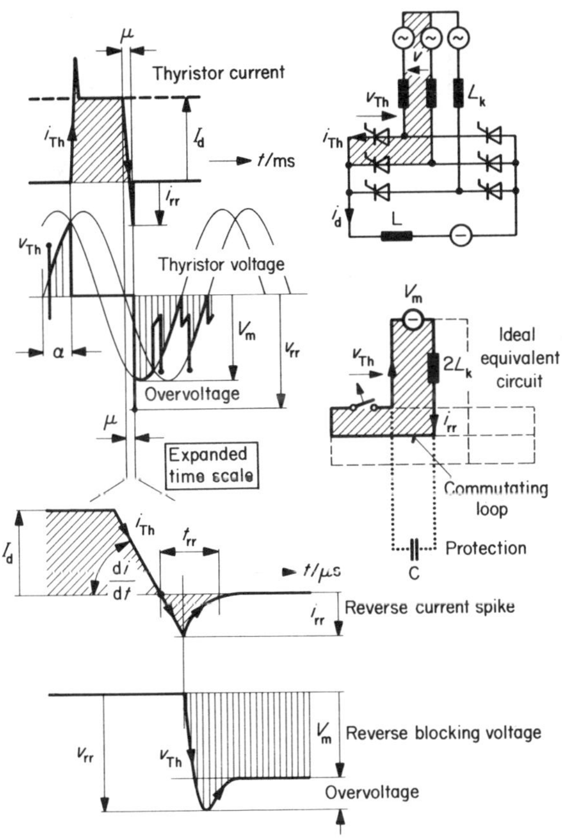

Figure 157. Actual thyristors (and diodes) do not block at the current zero but with a delay, and in so doing cut off a reverse current (i_{rr}) sharply. The valves are thereby subjected to overvoltages (v_{rr}) at the end of commutation (μ). Diagrammatic illustration to two time scales

switch, does not in reality break the circuit at a current zero, but while it is carrying current. The rapid interruption of current causes a rise in voltage in the circuit inductances, which appears as an *overvoltage* (v_{rr}) across the thyristors. This is illustrated in Figure 157.

The magnitude of the reverse current spike has a decisive influence on the resulting overvoltage. The reverse blocking current spike (i_{rr}) is primarily dependent upon the rate of fall (di/dt) of the load current. The greater di/dt, the higher is i_{rr} and the greater is the overvoltage. The maximum value of di/dt occurs approximately when $\alpha = 90°$ (so long as commutation takes place), the lowest value in discontinuous current operation. The magnitude of the reverse current spike is also affected by the load current, the temperature, and the type of thyristor. Even with thyristors of the same type, the spread among individual cells is relatively large.

The overvoltages (v_{rr}) developed across the turning-off thyristor at the end

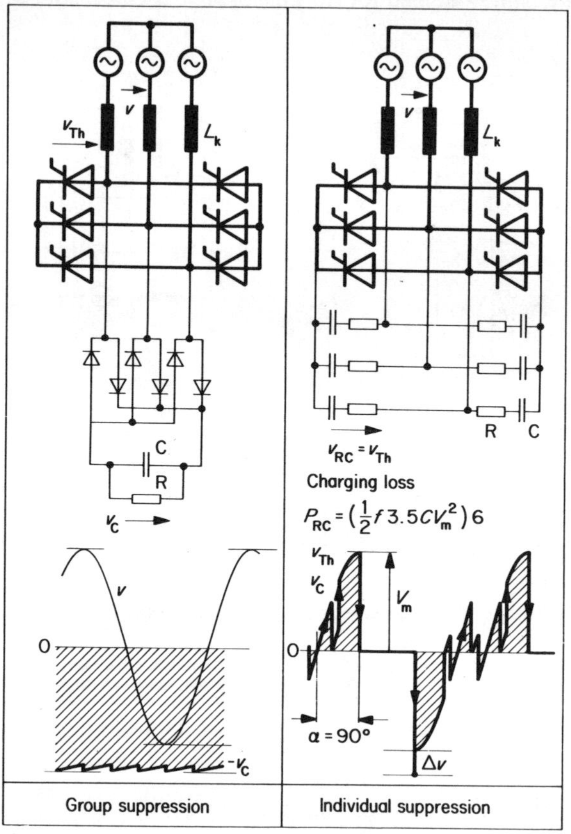

Figure 158. Overvoltages across the valves at the end of commutation (Figure 157) are most often damped by capacitors and resistors. Suppression is applied either to each thyristor individually (by 'snubbers') or to the complete converter. Suppression circuits and ideal voltage waveforms (neglecting oscillations due to transformer capacitance)

of the overlap period (μ)(Figure 157) can be many times the nominal voltage (V_m) and thus represent a risk to the thyristor. RC networks are usually employed as a means of protection in high-power converters.

The two principal overvoltage protection circuits for three-phase bridges are illustrated in Figure 158. Either each individual valve has a capacitor (C) and series resistor (R) connected in parallel with it (*individual suppression*) or a single capacitor (C) for the whole three-phase bridge is connected through a diode bridge, with a discharge resistor R (*group suppression*).

The functioning of the suppression can be simply explained as follows: due to the reverse current spike i_{rr} magnetic energy is stored in the commutating inductances L_K ($\frac{1}{2} L_K i_{rr}^2$). When the reverse current is terminated this energy is acquired by the capacitor ($\frac{1}{2} C v_C^2$) which is charged accordingly. The maximum voltage (v_{Cm}) on the capacitor is determined by the energy balance. The larger the capacitance C, the lower is the voltage v_{Cm}, which at the same time represents approximately the voltage stress on the thyristors. The energy stored in the capacitor C is then converted into heat in the resistor R. The capacitor is discharged and so the process can be repeated at the next commutation.

The suppression capacitor forms an oscillatory circuit with the commutating inductances. With group suppression, however, the capacitor is isolated from

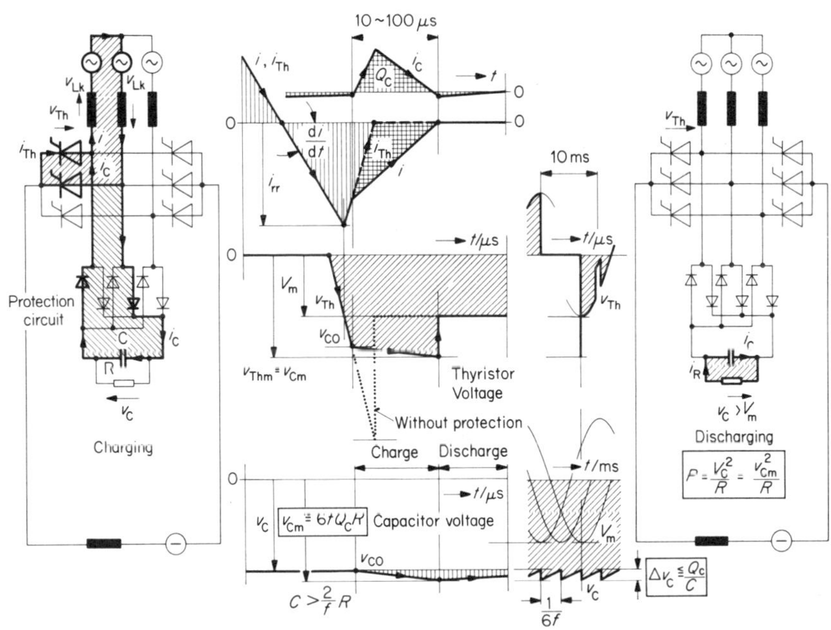

Figure 159. Simplified illustration of the operation of group suppression (Figure 158). Steady-state condition. The reverse current spike (i_{rr}) depends upon the rate of change of current, current loading, temperature, and thyristor type. The charge on the capacitor (Q_C) is affected by the type of thyristor, the rate of change of current, and the overvoltage ($v_{Cm} - V_m$), and is determined experimentally. Two time scales. Two circuit states. The energy from the a.c. supply system is only partly dissipated in the resistor R; some is dissipated in the thyristor itself

the three-phase system by the diodes, and the oscillatory process is in evidence mainly when the converter is switched on (Figure 155). With individual suppression, on the other hand, the oscillatory circuit is continually excited by the switching processes (commutations) in the converter and must therefore be damped. The resistance R in the individual suppression circuits is so chosen that the overshoot voltage across the thyristor (Δv) is as small as possible. the voltage steps across the capacitor in individual suppression (Figure 158) give rise to charging and discharging currents, which produce additional dissipation in the damping resistor (*charging losses*). The total loss is consequently greater with individual than with group suppression.

The functioning of *group suppression* is shown, idealized, in Figure 159. At the end of the commutating period the current in the thyristor passes through zero, with a constant rate of change (di/dt), into the negative direction, and reaches its peak i_{rr}. The reverse current then falls rapidly. Due to the collapse of current, the voltage (v_{Lk}) induced in the commutating inductances (L_k) increases the thyristor voltage (v_{Th}), which has risen in the meantime. When the thyristor voltage (v_{Th}) reaches the level of the capacitor voltage (v_{C0}), the capacitor begins to charge. The transformer current i is split into two components, capacitor current i_C and thyristor current i_{Th}. At the end of the process, the capacitor voltage, by virtue of a charge Q_C, reaches the level V_m, which also represents the maximum thyristor voltage v_{Th}. The diodes block and the capacitor is isolated from the converter. A voltage corresponding to the transformer line voltage appear across the thyristor. The capacitor discharges through the parallel resistor.

The overvoltages produced by commutation are periodic in nature; they are repeated for each thyristor once per cycle, and are therefore classified with the *continuous phenomena* of normal converter operation. The origin of *commutation overvoltages* is in the mode of operation of the converter valves themselves.

The turn-off behaviour of individual thyristors exhibits a statistical spread: the thyristors do not turn off simultaneously. In a string of *series-connected* thyristors (Figure 9), the first thyristor to block is subjected briefly to the full voltage of the whole string. To prevent this, the series-connected thyristors are provided with resistive-capacitive *voltage-sharing networks*. In addition, it is usual for series connection to choose thyristors with approximately equal stored charge.

Turn-off overvoltages

In an ideal current interruption the current is reduced suddenly to zero. In an inductive circuit *the instantaneous magnetic energy in the inductance* ($\frac{1}{2} Li^2$) *is thereby released and converted into electrostatic energy in the circuit capacitance* ($\frac{1}{2} Cv^2$). In a real situation part of the energy released is converted into heat in the arc of the switch and in the ohmic resistances. If the circuit capacitance is small, the energy exchange is accompanied by a high voltage.

In the case of current interruption by a switch or a fuse, the arc has physically the more important effect. However, the interruption process is very difficult to describe mathematically, because it is concerned with plasma behaviour *of a statistical nature.*

The following current-interruption processes are important in converters:

High-speed d.c. circuit breaker opening in the d.c. circuit
Fuse blowing in the valve circuit
Transformer disconnection in the three-phase a.c. circuit.

The ***high-speed d.c. circuit breaker*** serves to interrupt short-circuit currents (Figures 105 to 111). The interruption process in the case of a short circuit on an operating converter is illustrated in simplified form in Figure 160. When the short-circuit current (i_{dk}) reaches the tripping value of the breaker, the breaker contacts open and an electrical arc is formed. The ***arc voltage*** (v_{CB}) increases as a result of the contact movement and the increasing current. The maximum value of the arc voltage (v_{CB}) depends upon the construction of the circuit

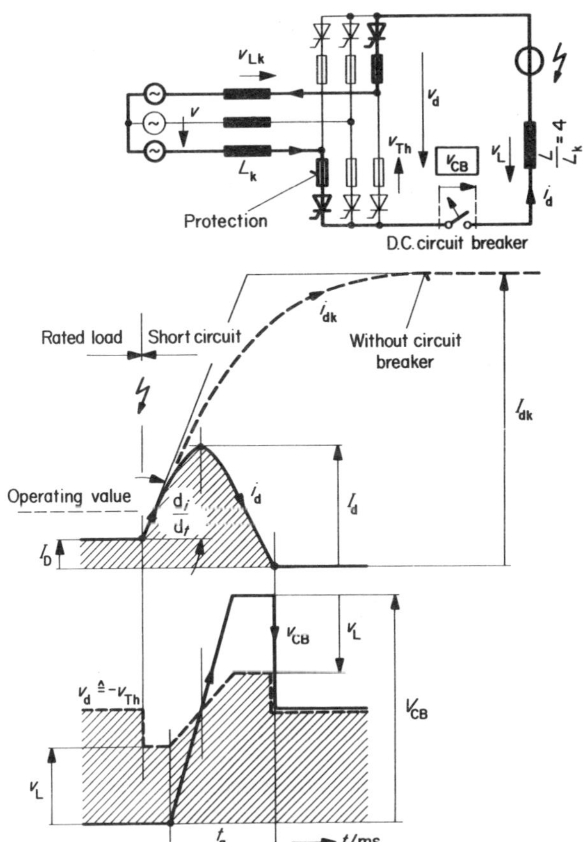

Figure 160. The interruption of short-circuit current by a d.c. circuit breaker produces a high arc voltage (v_{CB}) in the breaker. The d.c. inductance L supports part of the overvoltage and reduces the voltage stress on the thyristors

breaker and the circuit parameters. The arc voltage represents a back-voltage in the circuit, in consequence of which the short-circuit current falls. At the instant when the arc voltage is equal to the converter voltage ($v_{CB} = v_d$) the short-circuit current reaches its maximum (I_D). Following the collapse of current the arc is extinguished and the direct voltage v_d of the converter appears across the circuit breaker.

During the switching-off, the voltage across the thyristors is determined by the ratio of the circuit inductances ($2L_k/L$). With no d.c. circuit inductance ($L = 0$), the thyristors are subjected to the full arc voltage. Thus *the d.c.-side inductance reduces the voltage stressing of the valves during the current-interruption process in the d.c. circuit.* With increasing d.c. inductance (L, Figure 160) the short-circuit current (I_d) decreases, but the interruption time

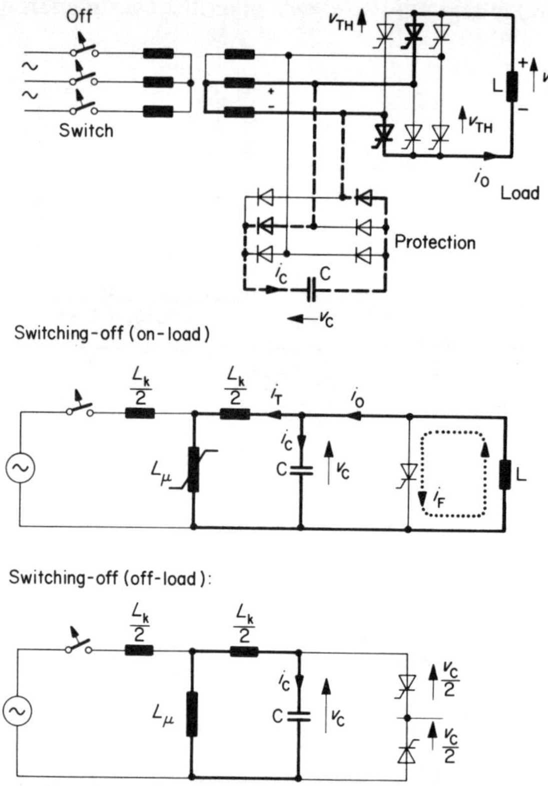

Figure 161. If the converter is switched off on the primary side of the transformer an overvoltage occurs as a result of the energy released from the no-load inductance L_μ of the transformer. Switching off on-load generally produces higher overvoltages than off-load, when, besides, the thyristors have to withstand only half the overvoltage. Equivalent circuit with single-phase transformer. The suppression capacitor is usually chosen only for the no-load switching condition, and on-load switching is not permissible

(t_a) becomes longer and the arc voltage (v_{CB}) higher. The current and voltage loading of the valves is, however, reduced. If the direct current is interrupted by the valve fuses, the relationships are similar to those that apply to switch opening. In this case there are two fuses in series in the short-circuit path.

In the event of a *fuse blowing* due to a valve breakdown (Figures 99 and 100) the parallel (Figure 9) or inverse-parallel-connected (Figures 122 and 123) thyristors are subjected to the arc voltage of the fuse.

In normal operating circumstances, the converter is usually switched off by an a.c. circuit breaker on the primary side of the transformer in the unloaded condition, with no current flowing. The converter is switched off in the blocking state. When *the unloaded transformer is switched off* the magnetizing current is interrupted and the magnetic energy released from the magnetizing inductance of the transformer ($\frac{1}{2} L_\mu i_\mu^2$) produces an increase in voltage across the energy-absorbing circuit capacitances.

As a protection against the overvoltages resulting from the switching-off of the transformer, a *capacitor* is normally connected to the secondary side of the transformer through diodes (Figure 161). The switching-off overvoltage developed across the suppression capacitor (v_C) is divided between two thyristors, assuming that the converter is blocking (Figure 161, bottom).

If *the transformer is switched off on the primary side while the converter is on load,* higher overvoltages are usually produced than on no-load. The two cases are illustrated for a single-phase case with reference to the equivalent circuit of the transformer in Figure 161. In switching off on-load, the released energy is greater than with no-load. The interrupted load current (i_0) is diverted to the magnetizing inductance of the transformer (L_μ) and causes an overvoltage. The transformer saturates. If the thyristors are triggered again during the switching-off, a freewheel path can be set up (i_F), as a result of which the overvoltage is eliminated.

In switching off on-load, the thyristors are subjected to the full voltage, in contrast to off-load switching.

Switching-off overvoltages generally exhibit a statistical variation. The magnitude of the overvoltage depends upon the instant of switching and upon the type of circuit breaker and its condition.

Thyristor stress due to high rate of rise of forward voltage

A steeply rising forward voltage (dv/dt) across a blocking thyristor can cause it to turn on (without a triggering pulse). The valve can be destroyed as a result.

The thyristors in the three-phase bridge can be affected by a steep-fronted surge from the a.c. or the d.c. side. Because of circuit inductances and capacitances (the thyristors themselves represent a capacitance in the blocking state; see Figure 6), a voltage step gives rise to oscillations. The rate of rise of voltage is then determined by the oscillation frequency.

Forward voltage steps can also appear across the thyristors in normal operation. The highest values are developed in inverter operation (Figures 64 and 121) with $\alpha = 120^\circ$. The reduction of unacceptably high dv/dt values can be effected by *thyristor suppression* (individual suppression; see Figure 158).

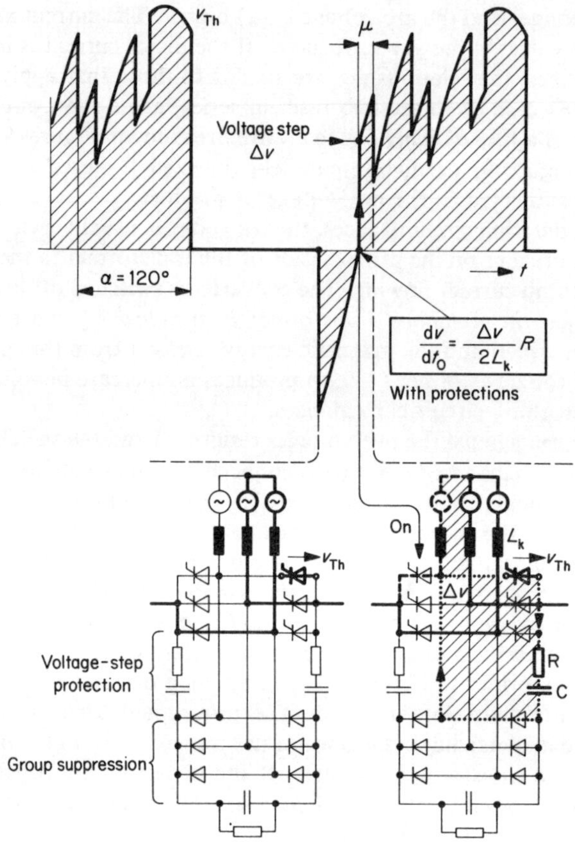

Figure 162. The unacceptably high rate of rise of forward voltage (dv/dt) on the thyristors, with overvoltage protection by group suppression (Figure 158), is reduced by two additional RC networks. Illustration of operation with $\alpha = 120°$ (representing the largest voltage step Δv)

With group suppression, additional RC circuits are incorporated as *step suppressors* (Figure 162).

Overvoltage protection devices

For overvoltage protection in converters, a *capacitor* with a series or parallel resistor is most frequently used (Figures 158 and 162). Non-linear devices based on silicon, selenium, or metal oxide are also employed. In the case of high-energy overvoltages, the converter, or the energy source, is short-circuited by a voltage-triggered *thyristor switch* (See Figure 170).

To prevent the turning-on of the thyristors by high voltages without triggering pulses, special *switching diodes* are also connected directly between the anode and the gate. Through the switching diode, the gate receives a triggering pulse in the event of an overvoltage. Turning-on by exceeding the breakover voltage is thus avoided.

MATHEMATICAL APPENDIX XXI

A capacitor, of capacitance C, damped by a series or parallel resistor R, forms the basis of most overvoltage protection circuits. The capacitances form oscillatory circuits in conjunction with the commutating, magnetizing, or load inductances (L_k, L_μ, L)(equations 14 to 17).

If the current i in the inductance L' is suddenly interrupted, the magnetic energy released is transferred to the capacitance. From the equality of energy,

$$\frac{1}{2} L' i^2 = \frac{1}{2} C \Delta v_C^2 \tag{139}$$

The increase in capacitor voltage is calculated as follows:

$$\Delta v_C^2 = v_{Cm}^2 - v_{C0}^2$$

where v_{C0} = initial voltage
v_{Cm} = maximum voltage

Hence,

$$v_{Cm} = \sqrt{v_{C0}^2 + i^2 \frac{L'}{C}} \tag{140}$$

The capacitance of the suppression capacitor is

$$C = \frac{i^2 L'}{\Delta v_C^2} \tag{141}$$

This equation can be applied to the case of the switching-off overvoltage (Figure 161) or to that of the commutation overvoltage (with individual suppression; see Figure 158).

As well as the *current step,* as assumed above, a *voltage step* also leads to an overvoltage. In an undamped oscillatory circuit (Figure 19) the capacitor voltage resulting from a voltage step E is given by

$$v_C = E(1 - \cos \omega_0 t) \tag{142}$$

where

$$\omega_0 = \frac{1}{\sqrt{L'C}} \tag{143}$$

The maximum capacitor voltage is therefore

$$v_{Cm} = 2E \tag{144}$$

The voltage peak v_{Cm} can be damped by resistors. When

$$R \geqslant \frac{1}{2} \sqrt{\frac{L'}{C}} \tag{145}$$

no periodic oscillations occur.

OVERCURRENT PROTECTION OF CONVERTER VALVES

Under normal operating conditions the temperature of the silicon wafer must not exceed a critical level (about 125 °C) if the valve is to maintain its control capability (Figure 154). Since the temperature of the silicon cannot be measured in operation, it is kept under the critical limit by proper current loading, according to previously determined current–time characteristics.

The maximum current limit is observed through ***electronic control and regulation***.

The thermal behaviour of the valve can, with some simplification, be compared with that of a ***thermal time-delay relay***.

The current rating of the converter, the choice of thyristor type, the number of valves connected in parallel, and the cooling arrangements are suited to the specified operating requirements.

High-speed d.c. circuit breakers and high-speed fuses are usually employed

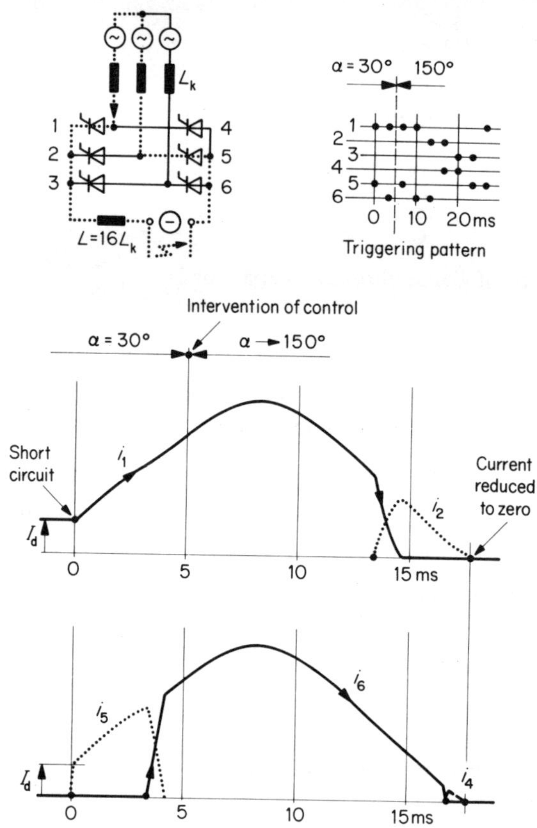

Figure 163. In the event of a d.c.-side short circuit on the three-phase bridge (Figures 105 and 106) the current can be cut off by control. Thyristor current waveform with $L/L_K = 16$. The triggering pulses are shifted suddenly to the position $\alpha = 150°$ after an assumed dead time of 5 ms. Example

for *short-circuit protection*. At the same time the short-circuit current is reduced by conrol. Current limiting by control is, however, often too slow.

The operation of the *high-speed d.c. circuit breaker* can be understood from Figure 160. By means of the d.c. circuit breaker the short-circuit current should be so limited that the valve fuses do not react (*discrimination*) and, obviously, so that the thyristors are not overstressed. The suppression of the short-circuit current by control is shown in the example of Figure 163. For the purposes of calculation it has been assumed that the electronic control takes effect only after a dead time of 5 ms, and that the triggering pulses are then shifted suddenly to the inverter limit of $\alpha = 150°$.

In the case of *loss of inverter control* (Figures 100, 102, and 108 to 110), however, electronic control is of little assistance, and the short-circuit current has to be interrupted by the high-speed d.c. breaker. Current-limiting by the *fuse* is illustrated in Figure 164 by reference to a typical characteristic (top).

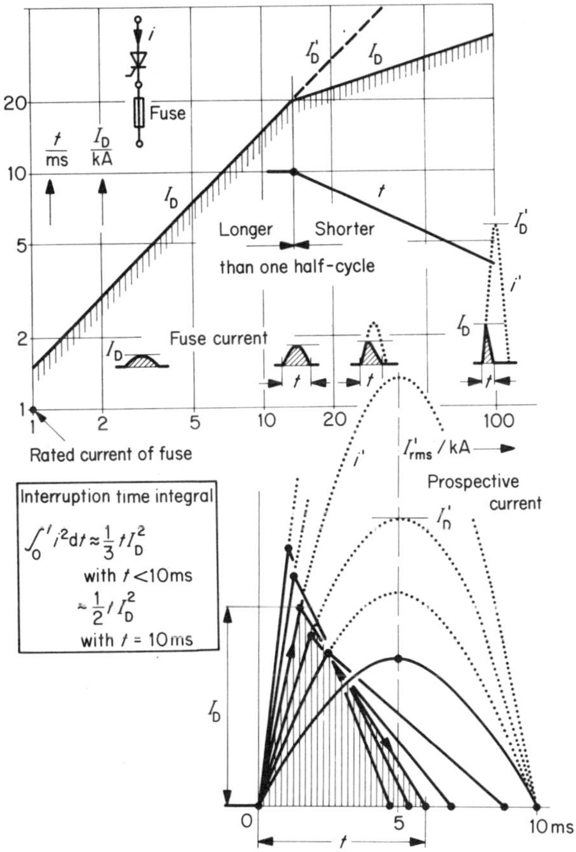

Figure 164. For interrupting short-circuit currents in coverters, d.c. circuit breakers and high-speed fuses (Figure 160) are normally used. Examples of fuse characteristics and forward current waveform (bottom). The main parameter is the fuse I^2t ($\int i^2\,dt$)(for short periods)

On the occurrence of a fault, the valve fuse must isolate the failed valve from the short-circuit current path (see Figures 99 and 100). The fuse associated with the short-circuit thyristor protects the other valves of the converter. In the failed valve mechanical destruction (*explosion*) is averted.

As a parameter for quantifying the interrupting capability of the fuse in the short-time range, the *I^2t let-through* (A^2s) is adopted. The smaller the I^2t value the more effectively the fuse protects. If the I^2t of the fuse is less than the *I^2t rating* of the thyristor, the valve is fully protected. I^2t values are also quoted for high-speed d.c. circuit breakers.

The problems of converter overload protection can be illustrated in a *current–time diagram* (Figure 165). The overload rating curve of the converter lies above the required operating load curve. The maximum current loading in operation is governed by the electronic control. In the long-time range the current loading is controlled by the thermal relay. The short-circuit current is

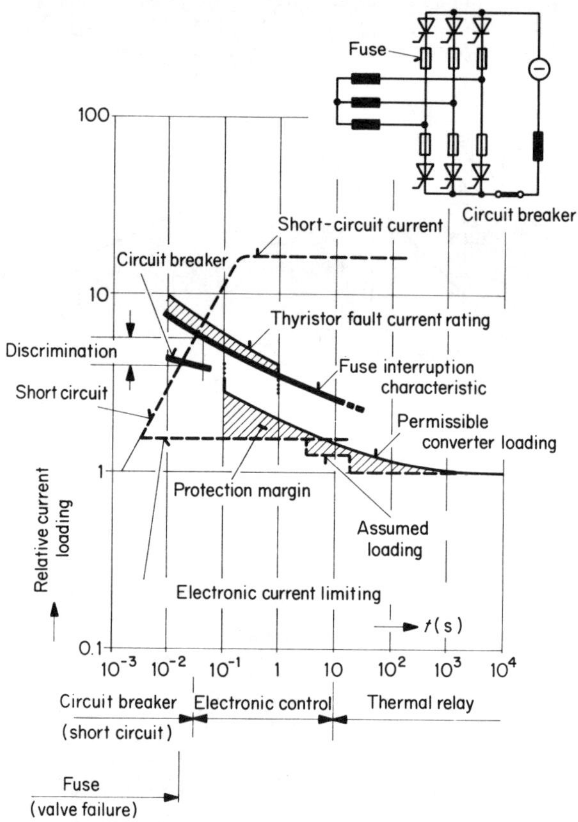

Figure 165. The current loading characteristic of the converter must lie above the level of the load duty cycle. The interruption characteristics of the fuses and the high-speed d.c. circuit breaker are normally coordinated selectively and are below the current rating curve of the thyristors. Example of converter overcurrent protection

limited by the d.c. circuit breaker. In the event of a valve breakdown, the valve fuse disconnects the defective thyristor without permitting the current rating limit of the thyristors to be exceeded (in fact it is never reached). The thyristor current rating is determined by the thermal limit of the valve (Figure 154). Following the overload, the thyristor must be free of voltage stress until it has cooled, and is in the meantime not controllable.

MATHEMATICAL APPENDIX XXII

For the assessment of the short-circuit protection means for times in the region of one cycle, the I^2t let-through

$$\int_0^t i^2 \, dt$$

of the thyristors and diodes is used as a measure of their heat absorption capability, where

i = instantaneous valve current
t = duration of load

The valve may be considered as protected when the I^2t let-through of the protection device is less than the I^2t rating of the valve.

The currents interrupted by circuit breakers and fuses exhibit a roughly triangular waveform (Figure 164). In this case the I^2t can be calculated as

$$\int_0^t i^2 \, dt = \frac{1}{3} t I_F^2$$

where I_F = peak value of current

For a half-sine-wave,

$$\int_0^t i^2 \, dt = \frac{1}{2} t I_F^2$$

15

Thermal Characteristics and Cooling of Converter Valves

LOSSES AND THERMAL RESISTANCE

All valve parameters, such as forward and reverse blocking voltages, forward voltage drop, reverse recovery current, etc. (Figure 5), are dependent upon the temperature of the semiconductor material (silicon). To ensure normal operation of the thyristors (and diodes), therefore, the temperature of the silicon wafer (Figure 4) must be kept within certain limits (Figure 154). If the *temperature limit* is exceeded the valve may be destroyed.

Electrical losses occur in the silicon wafer in all the periods of the operating cycle of the valve (Figure 6); in converters operating on the three-phase mains, however, the *conduction loss* of the valves is the most significant.

For the calculation of the conduction loss the converter valve is represented by an equivalent resistance (r) and a threshold voltage (v_0) (Figures 6 and 166). The values in this equivalent circuit are obtained from the forward conducting characteristic (Figure 166, top).

The thermal capacity and the surface area of the silicon wafer are very small in relation to the losses that are developed, and without an additional *cooler* (Figures 4 and 8) the utilization of thyristors and diodes would be very poor. In practice, therefore, converter valves are invariably used with coolers.

If the valve is loaded by a constant current (constant loss $= P$) and the temperature rise of the silicon wafer is measured, an exponentially increasing curve is obtained. After a certain time the steady-state condition is reached and the temperature does not rise any further. The height of the heating curve is proportional to the loss. The converter valve behaves like an impedance in an electrical circuit. In the steady state the thermal characteristic of the valve can be compared to that of an ohmic resistance in an electrical circuit.

Between thermal and electrical circuits there exist the following analogies:

Power loss P (W):	Current i (A)
Temperature rise θ (°C):	Voltage V (V)
Thermal resistance R_{th} (°C/W):	Resistance R (Ω)
Thermal capacity C_{th} (Ws/°C):	Capacitance C (F)
Time constant τ (s):	Time constant τ (s)

The equations for a thermal circuit are similar in form to those for an electrical circuit. The thermal characteristics of a converter valve can be represented by an equivalent circuit with simple RC elements (Figure 167). To a first approximation the parameters of the equivalent circuit can be deduced graphically from the heating and cooling curve (the thermal impedance curve, or *transient thermal resistance*), as shown in Figure 167. With the aid of the thermal equivalent circuit the valve temperature can be calculated for any loading curve.

A simpler method of estimating converter valve temperature is illustrated in Figure 168. The given power pulses of the valve loading curve are regarded as being made up of a series of constant positive and negative pulses (the *superposition principle)*. For each of the individual constant loads a temperature excursion is plotted from the curve of transient thermal resistance (right). The addition of these gives the required temperature excursion (bottom). (The transient thermal resistance curve—Figure 167—is mainly used for calculating intermittent load cycles or surge loading.)

The permissible continuous current for the converter valve is determined from the permissible losses (P)(Figure 166). The permissible losses in the valve are derived from the quotient of the permissible temperature rise (θ) and the thermal resistance (R_{th}).

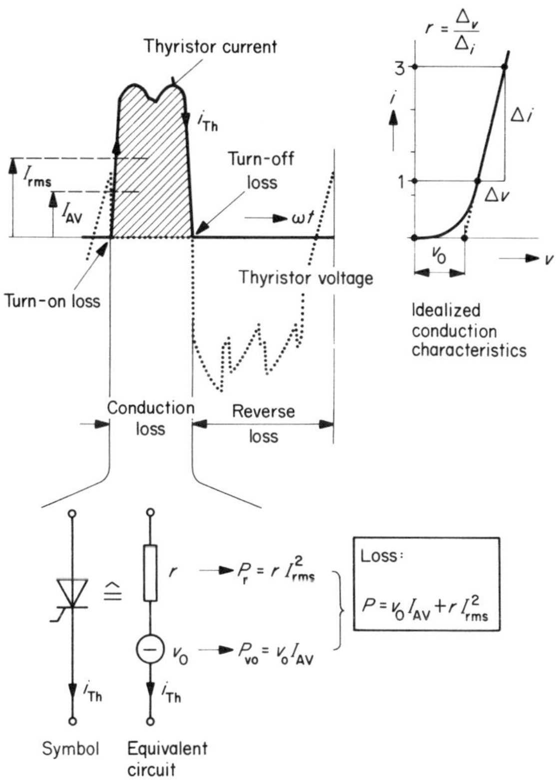

Figure 166. In the course of the switching cycle in the operation of the converter valve, losses occur in the silicon wafer. The most significant part is the forward conduction loss, which can be simply calculated by means of the equivalent circuit (r, v_0). The parameters of the equivalent circuit are derived from the idealized forward characteristic

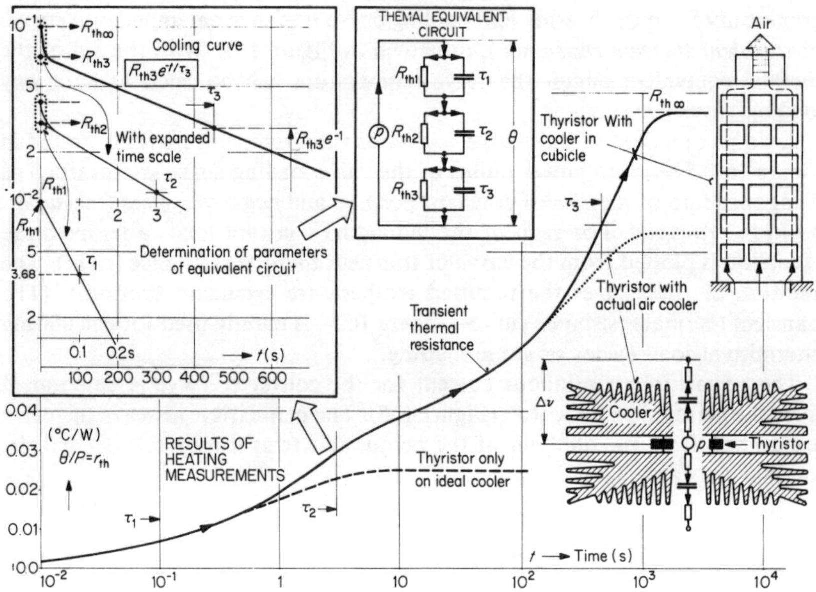

Figure 167. The temperature curve of the silicon wafer (θ) with constant current loading (loss) P leads to the curve of transient thermal resistance (r_{th}). From analysis of the curve a thermal equivalent circuit for the valve can be deduced, whereby thermal calculations can be carried out in the same form as in electrical circuits

COOLING

The heat produced by the losses in the silicon wafer of the valve flows through the material of the cooler to its surface. There it is absorbed by the cooling medium (gas or liquid) and removed. The larger the surface area of the cooler and the greater the flow of the cooling medium (turbulent flow is aimed at) the more effective is the cooler. Also important is the thermal resistance of the cooler material and especially the contact thermal resistance between the mating surfaces.

Natural air cooling is used in converters at low powers—forced air cooling, with a fan, at higher powers. For very high powers or in special applications cooling by water, SF_6 (sulphur hexafluoride), freon, or oil is preferred.

MATHEMATICAL APPENDIX XXIII

The mean power loss in the converter valve is given by (Figure 166)

$$P = I_{AV}v_0 + rI_{rms}^2 \quad (W) \tag{146}$$

where I_{AV} = mean value of valve current (A)
I_{rms} = r.m.s. value of valve current (A)
v_0 = threshold voltage of idealized characteristic (V)
r = equivalent (incremental) resistance (Ω)

or, more simply,

$$P = I_{AV}v_0 + k_f^2 rI_{AV}^2 \tag{147}$$

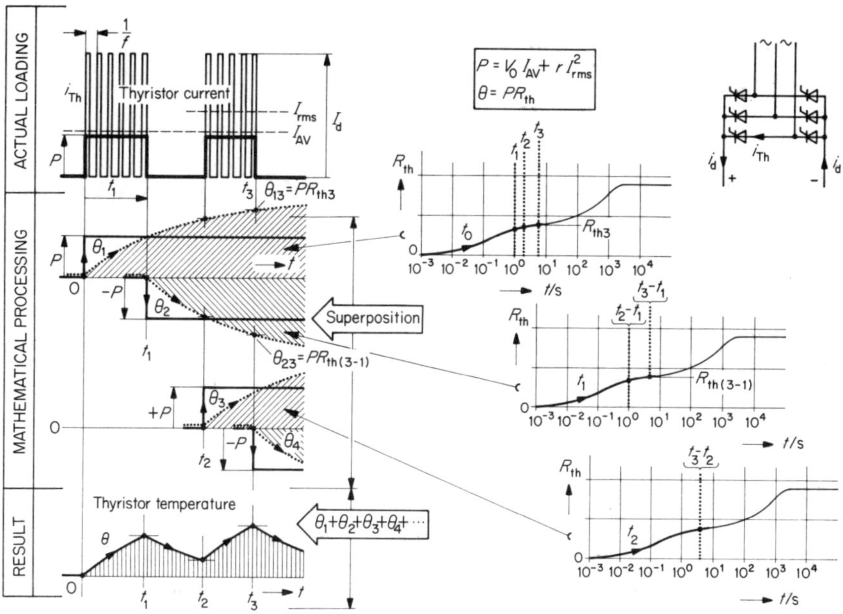

Figure 168. The transient thermal resistance curve of the converter valve (Figure 167) facilitates the estimation of valve temperature under surge conditions and in intermittent operation. The superposition principle can usefully be applied to this (treating the dissipation in terms of positive and negative power blocks). Example of the application of the method. Estimation of the thyristor temperature after the second power block at time t_3. f = supply frequency

where $k_f = I_{rms}/I_{AV}$ = form factor ($\approx \sqrt{3}$ for the three-phase bridge)

Across the thermal resistance R_{th} of the valve (Figures 167 and 168) the power flow P produces a temperature rise

$$\Delta\theta = PR_{th} \tag{148}$$

The permissible dissipation is derived from the maximum permissible temperature rise

$$P_{max} = \frac{\Delta\theta_{max}}{R_{th}} \tag{149}$$

From equations (149) and (147), the permissible mean current in the valve is

$$I_{AVmax} = \frac{\sqrt{v_0^2 + 4k_f^2 r P_{max}} - v_0}{2k_f^2 r} \tag{150}$$

In the air channel (Figure 167) the temperature of the air flowing over the cooler is raised by the valve losses through approximately

$$\Delta\theta_A \approx \frac{P}{1.2Q} \quad (^\circ\text{C}) \tag{151}$$

where Q = rate of flow of cooling air (l/s)

16

Specification of Converters

CALCULATION SEQUENCE

During the planning stage the *installation scheme* and the *main installation data* are first established. In the case of drives, the type of machine—d.c., induction, or synchronous motor—must be chosen (Figure 119) and the converter circuit decided upon—six-pulse or twelve-pulse converter, simple or sequential control, inverse-parallel or H-circuit (Figures 118, 123, 143, and 144).

In a *d.c. drive* (Figure 120, for example) the motor current (I_m) and the motor voltage (V_m) form the starting point for the dimensioning of the converter. Since the converter parameters—the ideal no-load direct voltage (V_{dio}) and the direct current (I_d)—are determined by the converter transformer, the *transformer data*—the secondary voltage and current (V, I)—

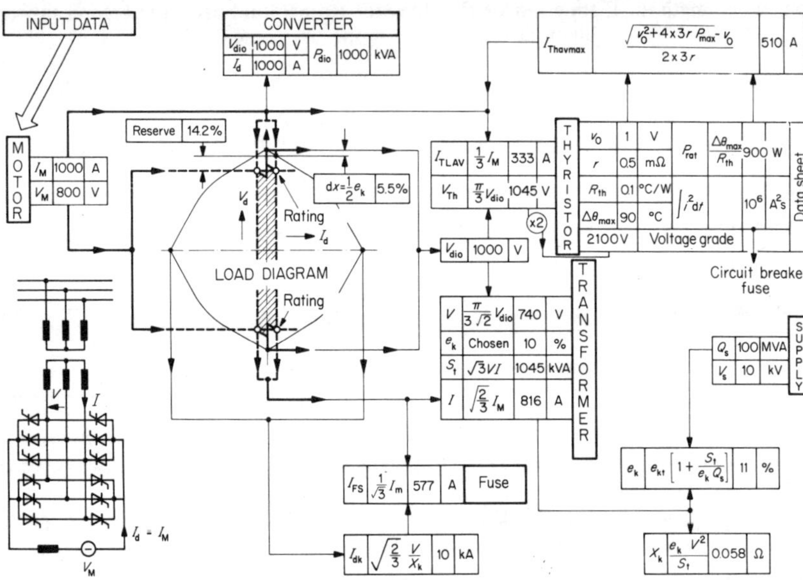

Figure 169. The basis for the specification of the converter is the load diagram (Figures 56 and 63). Simplified presentation of the calculation sequence with a numerical example of a reversing d.c. drive. The voltage margin allows for ohmic resistances, supply voltage fluctuations, etc. In the case illustrated a transformer rating of 1045 kVA is required for a motor rating of 800 kW

are calculated as a first step. The basis of calculation is the *load diagram* of the three-phase bridge (Figures 56 and 63). The operating point in the converter load diagram is chosen with reference to the short-circuit level of the supply (Q_s in Figures 116 and 117), additional ohmic resistance voltage-drops in the three-phase system (Figure 112), and in the d.c. circuit (Figure 113), certain triggering of the thyristors (Figure 147), supply-voltage variations, etc. The transformer voltage is so chosen that the rated motor voltage (and therefore the rated motor speed) is ensured under the most adverse operating conditions.

From the transformer voltage (V) the *voltage loading of the thyristors* (V_m in Figures 57 and 64) is obtained. Normally thyristors are rated with a *voltage safety factor of two* (Figure 154).

For higher voltages either the valves (Figure 9) or the bridges (Figures 141 to 144) are connected in *series.*

The motor current ($I_d = I_m$) leads to the *current loading* (I_{ThAV}) *of the thyristors* (Figure 58). Thermal calculations (Figures 166 to 168) decide whether the losses in the proposed type of thyristor, with the assumed load cycle and converter construction, will result in an acceptable temperature rise. For higher currents thyristors are connected in parallel (Figure 9).

Figure 169 illustrates in simplified form the calculation sequence in the specification of a reversing d.c. drive. The quantitative relationships are shown approximately by the numerical example.

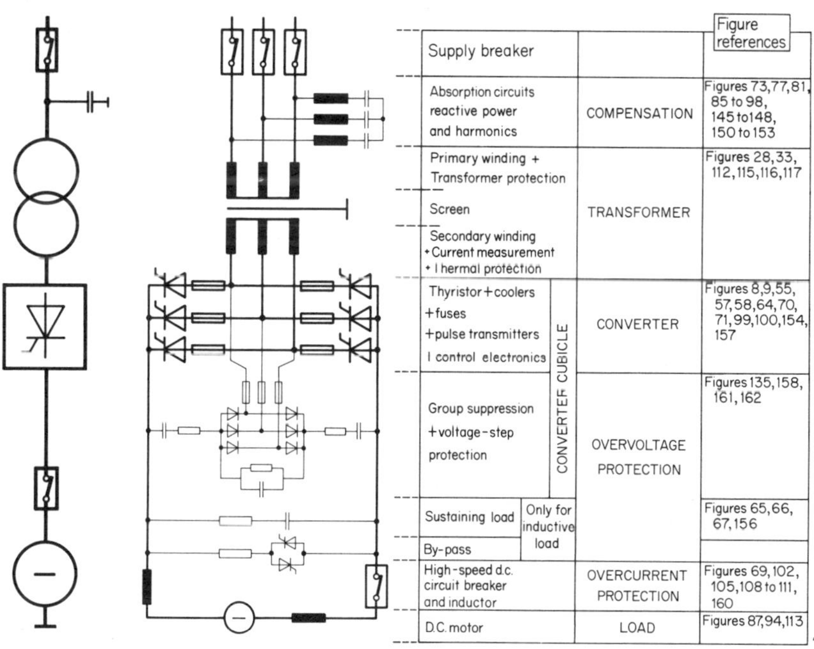

Figure 170. The choice of converter protection devices is affected by a whole series of economic and technical factors. Representation of a converter d.c. drive with a summary of typical overvoltage and overcurrent protection devices

The converter specification also embraces the choice and dimensioning of the *protection devices*. Overvoltage and overcurrent protection is summarized in Figure 170, in which the whole converter installation is represented.

An additional d.c. inductor is included, either air-cored, within the framework of the protection philosophy (Figures 105 to 111, 160, and 163), or iron-cored for smoothing the direct current (Figure 69).

CHOICE OF CONVERTER CIRCUIT

The fully controlled *three-phase bridge,* equipped with thyristors, has established itself as the most important converter circuit for three-phase systems. The operation of the converter does not require a neutral connection to the three-phase system, which facilitates direct connection to the supply. If necessary, a simple three-phase matching transformer can be used.

The fully controlled three-phase bridge can function as a rectifier or as an inverter (Figure 64). If only rectifier operation is required, one commutating group of the circuit can be furnished with diodes (forming a *half-controlled three-phase bridge*), or a diode bridge can be controlled on the a.c. side by inverse parallel-connected thyristors (an *a.c. regulator*).

Exactly equivalent to the three-phase bridge is the *double-star circuit with interphase reactor*. The interphase-reactor circuit was the principal circuit used

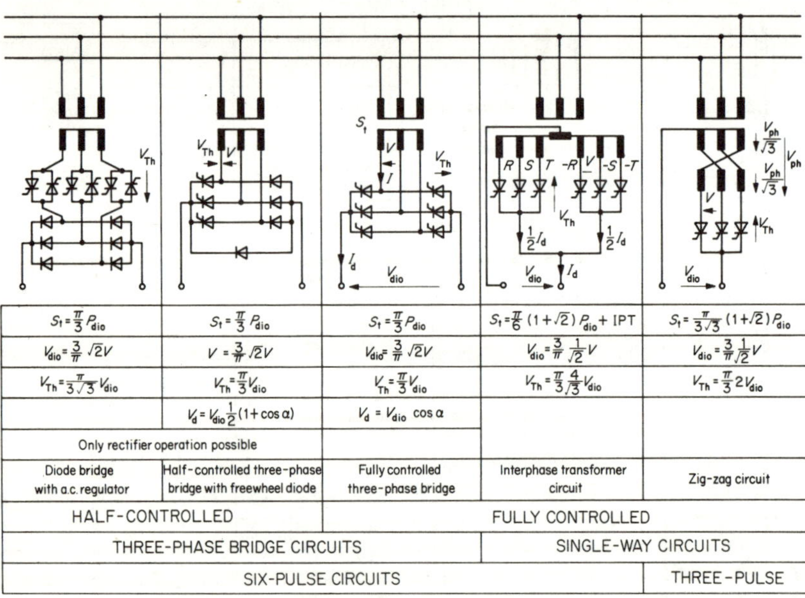

Diode bridge with a.c. regulator	Half-controlled three-phase bridge with freewheel diode	Fully controlled three-phase bridge	Interphase transformer circuit	Zig-zag circuit
$S_t = \frac{\pi}{3} P_{dio}$	$S_t = \frac{\pi}{3} P_{dio}$	$S_t = \frac{\pi}{3} P_{dio}$	$S_t = \frac{\pi}{6}(1+\sqrt{2}) P_{dio} + \text{IPT}$	$S_t = \frac{\pi}{3\sqrt{3}}(1+\sqrt{2}) P_{dio}$
$V_{dio} = \frac{3}{\pi}\sqrt{2}V$	$V = \frac{3}{\pi}\sqrt{2}V$	$V_{dio} = \frac{3}{\pi}\sqrt{2}V$	$V_{dio} = \frac{3}{\pi}\frac{1}{\sqrt{2}}V$	$V_{dio} = \frac{3}{\pi}\frac{1}{\sqrt{2}}V$
$V_{Th} = \frac{\pi}{3\sqrt{3}} V_{dio}$	$V_{Th} = \frac{\pi}{3} V_{dio}$	$V_{Th} = \frac{\pi}{3} V_{dio}$	$V_{Th} = \frac{\pi}{3}\frac{4}{\sqrt{3}} V_{dio}$	$V_{Th} = \frac{\pi}{3} 2 V_{dio}$
	$V_d = V_{dio}\frac{1}{2}(1+\cos\alpha)$	$V_d = V_{dio}\cos\alpha$		
Only rectifier operation possible				
HALF-CONTROLLED		FULLY CONTROLLED		
THREE-PHASE BRIDGE CIRCUITS			SINGLE-WAY CIRCUITS	
SIX-PULSE CIRCUITS				THREE-PULSE

Figure 171. The full controlled three-phase bridge is the most frequently used on three-phase supply systems. If only rectifier operation is required, a more economical circuit (a.c. regulator with diode bridge or half-controlled bridge) can be used. In the low-voltage range the single-way circuits can be advantageous. For higher powers several bridges are connected in series or in parallel (for twelve- or twenty-four pulse operation)

with mercury arc rectifiers (Figure 3). The transformer for the double-star circuit has to be rated for a higher power than that for the three-phase bridge, and an additional (interphase) reactor is necessary. In voltage terms the valves in the double-star circuit have twice the loading (for the same direct voltage) of those in the bridge, but at the same time the direct current obtained is twice as great. For the same d.c. power, therefore, the number of thyristors required is theoretically the same. For very high currents and low voltages, however, the double-star circuit is in practice more economical than the three-phase bridge.

For low powers the *three-pulse* circuit can also be used. The secondary windings of the transformer are then arranged in a zig-zag connection. The relative utilization of the rated power of the transformer is, however, poorer by about one-third, and the ripple in the direct current is greater.

The three-phase converter circuits referred to are summarized in Figure 171.

THE MAIN CHARACTERISTICS OF CONVERTERS IN BRIEF

The converter can be considered and explained from various points of view, so that quite different models present themselves according to the operating circumstances (Figure 172).

The converter exhibits some remarkable properties:

1. With a purely resistive load it presents an inductive load to the supply (Figure 21).

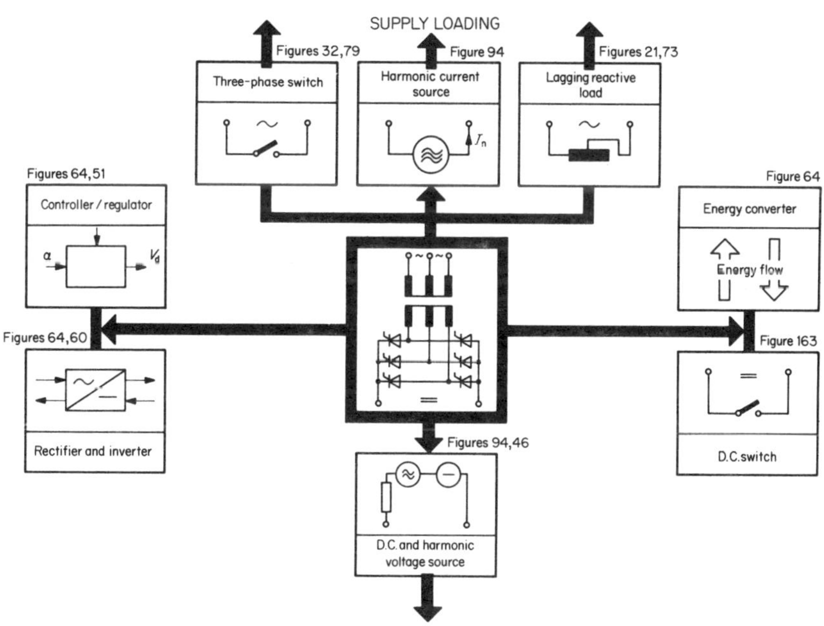

Figure 172. The converter can be variously represented according to the operating circumstances: as rectifier and inverter, as a regulator, energy-direction controller, d.c. or a.c. switch, harmonic and reactive power source, etc. Diagrammatic representation. Figure references

2. With a purely inductive load, under transient conditions, the direct current exhibits a time constant (Figures 31 and 39).
3. Under steady-state conditions the circuit state varies continuously (Figures 31 and 39).
4. With completely smooth direct current the converter functions as a harmonic generator to the supply (Figure 94).

Bibliography

There exists a whole series of excellent handbooks that treat the subject of power electronics from various points of view. We list here only those English language books that have been published in recent years by John Wiley and Sons:

E. W. Kimbark: Direct Current Transmission, Vol. 1, 1971.

L. Gyugyi and B. R. Pelly: Static Power Frequency Changers, 1976.

B. M. Bird and K. G. King: An Introduction to Power Electronics, 1983.

B. R. Pelly: Thyristor Phase-Controlled Converters and Cycloconverters, 1971.

Index

All the entries in this index refer to figure numbers (not to page numbers), since the most important information is contained in the figures.